Safety-II in Practice

Safety-I is defined as the freedom from unacceptable harm. The purpose of traditional safety management is therefore to find ways to ensure this 'freedom'. But as socio-technical systems steadily have become larger and less tractable, this has become harder to do. Resilience engineering pointed out from the very beginning that resilient performance – an organisation's ability to function as required under expected and unexpected conditions alike – needed more than the prevention of incidents and accidents. This developed into a new interpretation of safety (Safety-II) and consequently a new form of safety management.

Safety-II changes safety management from protective safety and a focus on how things can go wrong to productive safety and a focus on how things can go well. For Safety-II, the aim is not just the elimination of hazards and the prevention of failures and malfunctions but also how best to develop an organisation's potentials for resilient performance – the way it responds, monitors, learns, and anticipates. That requires models and methods that go beyond the Safety-I toolbox. This book introduces a comprehensive approach for the management of Safety-II called the Resilience Assessment Grid (RAG). It explains the principles of the RAG and how it can be used to develop the resilience potentials. The RAG provides four sets of diagnostic and formative questions that can be tailored to any organisation. The questions are based on the principles of resilience engineering and backed by practical experience from several domains.

Safety-II in Practice is for both the safety professional and the academic reader. For the professional, it presents a workable method (RAG) for the management of Safety-II, with a proven track record. For academic and student readers, this book is a concise and practical presentation of resilience engineering.

Erik Hollnagel (PhD, psychology) is Professor at the Department of Regional Health Research, University of Southern Denmark and Chief Consultant at the Centre for Quality, Region of Southern Denmark. Erik is also Adjunct Professor, Central Queensland University, Australia; Visiting Professorial Fellow, Macquarie University, Australia; Visiting Fellow, Institute for Advanced Study, Technische Universität München (Germany); and Professor Emeritus at École des Mines de Paris (France) and the University of Linköping (Sweden). Since 1971, he has worked at universities, research centres, and industries in several countries and with problems from many domains, including nuclear power generation, aerospace and aviation, air traffic management, software engineering, healthcare, and land-based traffic. His professional interests include industrial safety, human factors, resilience engineering, systems theory, and functional modelling. He has published more than 350 papers and authored or edited 24 books.

Safety-II in Practice

Developing the Resilience Potentials

Erik Hollnagel

Routledge
Taylor & Francis Group

LONDON AND NEW YORK

First published 2018
by Routledge
2 Park Square, Milton Park, Abingdon, Oxon OX14 4RN

and by Routledge
711 Third Avenue, New York, NY 10017

Routledge is an imprint of the Taylor & Francis Group, an informa business

British Library Cataloguing-in-Publication Data
A catalogue record for this book is available from the British Library

Library of Congress Cataloging-in-Publication Data
A catalog record for this title has been requested

ISBN: 978-1-138-70891-4 (hbk)
ISBN: 978-1-138-70892-1 (pbk)
ISBN: 978-1-315-20102-3 (ebk)

Typeset in Garamond
by codeMantra

To my beloved wife Agnes

Contents

Figures and tables

Figures

Tables

Preface

Like a child, resilience engineering has grown, although it has not yet come of age. It also has a date – or at least a year – of birth and possibly even a place. The first gathering of experts to discuss resilience engineering was in October 2004, in the Swedish town of Söderköping. This means that resilience engineering at the time of writing (2016) is 12 years of age. The gestation period was, however, rather long. The first noted use of the term was a presentation that David Woods made in 2000 for NASA as it considered how better to manage risky missions following a series of space exploration accidents (Woods, 2000). In parallel to that Hollnagel (2001) began to explore safety as a balance – or imbalance – in a key trade-off between efficiency and thoroughness, and this and other fundamental trade-offs have provided some of the theoretical foundations for resilience engineering (Hollnagel, 2009a). The development of resilience engineering from 2004 onwards has been documented in five books as well as numerous conference presentations and journal papers. More recently, the specific application of resilience engineering principles to healthcare has become a field of activity on its own – called Resilient Health Care.

The practical interest in resilience engineering was noticeable from the very beginning. Part of the motivation for developing this new field of enquiry was a growing dissatisfaction, if not outright frustration, with the established approaches to safety analyses and safety management. Since safety generally was defined as the freedom from unacceptable harm, or words to that effect, the purpose of safety management was naturally to ensure this 'freedom'. But as sociotechnical systems continued to become larger and less tractable, the much desired 'freedom' became harder to obtain. Resilience engineering recognised from the very beginning that it was necessary not only to prevent incidents and accidents but also to ensure resilience – defined as an organisation's ability to function as required under expected and unexpected conditions alike. Resilience engineering thereby offered a new interpretation of safety management.

The difference between the two perspectives was accentuated by the introduction of the terms 'Safety-I' and 'Safety-II' as a way to clarify the purpose of safety management in today's world. Where a Safety-I perspective emphasises protective safety and thereby a focus on how things can go wrong, a Safety-II perspective emphasises productive safety and a corresponding focus on how things can go well. Although focusing on how acceptable outcomes come about and looking for ways to support them is neither new nor exotic, there were few concepts or methods in Safety-I that could contribute to do so.

The purpose of this book is to provide concepts and methods that can be used to manage Safety-II, or in other words concepts and methods that can be used to improve how an organisation functions as a whole and not just with regard to safety seen as the freedom from risk and harm. Chapters 1 and 2 provide a brief, general introduction to safety management and to resilience engineering. Chapter 3 discusses the nature of resilient performance and introduces the concept of resilience potentials. Chapters 4 and 5 describe the details of the resilience potentials and how they can be assessed. Since it is crucial to understand how the resilience potentials function as a whole, Chapter 6 introduces a functional model of resilient performance. Based on this, Chapter 7 outlines an overall strategy to manage an organisation's performance and develop its resilience potentials. Finally, Chapter 8 offers some thoughts on the changing face of safety and a hint of the way forward.

Chapter 1

Safety management anno 2016

Safety management has a brief but chequered history where the institutionalised concern for safety at places of work, in the sense of efforts to prevent harm to people, goes back about 200 years. The initial safety concerns focused on the harm and injuries that could befall people who were at work. This was understandable considering the nature of work, not least the nature of the relatively unsophisticated technology that was used. Seen from the perspective of industrial work in the second decade of the 21st century, the technology of the workplace in the 19th century was quite simple, not least because the level of automation was low. Work processes were also relatively independent of each other and would typically show the linear dependency of the assembly line. All this changed dramatically around the middle of the 20th century, not least because of the advent of new technologies and sciences: digital computers, telecommunication, cybernetics, and information theory. Technology became more powerful but also more complex, processes became more integrated and dependent on each other, customer demands to quality and reliability grew, and the pace of work increased relentlessly. Safety was no longer limited to the prevention of injuries to the people at work, but had to consider the possible hazards of the technology being used to customers, to innocent bystanders, and to society.

The development of many of the new technologies had been, to a large extent, driven by the needs of the military during WW-II and its continuation in the Cold War that followed. The precursor to industrial safety management systems (SMS) was the concern for system safety engineering that began in the 1950s in the U.S. Air Force Ballistic Missile Division. The increased complexity of the equipment created a need to ensure that technology would function as intended with optimum safety within the constraints of operational effectiveness. The same need could soon be found in the civilian sector where complex technologies were enthusiastically greeted as a way to provide better products and services to the consumer as well as higher profitability. Despite the many problems that this has led to, with regard to safety as well as other aspects, these needs have so far shown little sign of weakening.

SMSs started to appear around the beginning of the current millennium and a standard for SMS was arguably introduced by the International Civil Aviation Organisation (ICAO) and described as follows:

> A safety management system (SMS) is an organized approach to managing safety, including the necessary organizational structures, accountabilities, policies and procedures. (ICAO, 2006)

ICAO's SMS standard was, quite typically, motivated by a growing number of serious aviation accidents. In that sense, it was no different from the safety legislation in the 19th century or from practically every other effort or initiative to improve safety. Indeed, ICAO's *Safety Management Manual* (page 2–1) defines safety as:

> … the state in which the possibility of harm to persons or of property damage is reduced to, and maintained at or below, an acceptable level through a continuing process of hazard identification and risk management.

Across all industries and professions, safety is associated with being *free from* harm and therefore being *without* the consequences of failure, damage, accidents, or other nondesirable events. The purpose of safety management is consequently to ensure that the number of things that go wrong, i.e., the hazards, or the number of adverse outcomes is as low as possible – with zero harm as the ideal. However, as a growing number of industries and practitioners have realised, this is not enough in the world of today. Identification and elimination of hazards, and prevention of and protection against unwanted outcomes, is inadequate for nontrivial sociotechnical systems, as resilience engineering has pointed out. A state of safety must also include looking at the things that go well in order to find ways to support and facilitate them. This goes for the individuals at work as well as for the organisation and the overall SMS.

Managing what is not there

The two most serious problems with the common approach to safety relate to the way that safety is measured and to the way that safety is studied.

The measurement problem is simply that an *increase* in safety is represented by a *decrease* in what is measured. Thus, a *lower* number of reported accidents (or other unwanted outcomes) is seen as representing a *higher* level of safety. The purpose of safety management is continuously to reduce or eliminate adverse outcomes and thereby achieve the enviable state of 'freedom from harm'. But it is only possible to know how well an SMS works if there is something to measure. Therefore, the better the job an SMS does the less information there is about how to make improvements. This corresponds to the well-known *regulator paradox*, where the absence of feedback ultimately leads to a loss of control

(Weinberg and Weinberg, 1979). The essence of the paradox is that the task of a regulator is to eliminate variation but this variation is the ultimate source of information about how well the regulator works. Therefore, the better the job a regulator does the less information it gets about how to improve. Thus, if an investment in safety does not lead to measurable results, such as a reduction in the number of accidents, then there is no way of knowing whether the investment had the desired effect. Furthermore, if the number of accidents is low to begin with, it is unreasonable to expect that the effect of improvements made can ever be measured.

The problem of how safety is studied comes, in a way, from the very definition of safety as the (relative) absence of injury or harm (see also Chapter 8). Situations where such harm occurs are said to represent a lack of safety or to be due to the absence of safety. It is therefore nothing less than paradoxical that we try to improve our understanding of safety by studying situations where we acknowledge there is a lack of safety. Safety science thus differs from other sciences by trying to study its subject matter in situations where it is absent, rather than in situations where it is present. It is little wonder that progress has been so slow.

It is widely believed that safety management should pursue the 'zero vision' and that the level of risk should be As Low As Reasonably Practicable (ALARP). Although it makes intuitive sense to be free from incidents and accidents, it does not make much sense that the goal of safety management is to be *without* something. It stands to reason that it is difficult to manage something that is not there, that it is difficult to measure it, and that it is difficult to understand it. It is of limited comfort that safety is not alone in doing that. The same problem can be found in Statistical Process Control, in lean manufacturing, and in Total Quality Management.

Safety management: a focus on details

When we try to explain something, in particular events that in one way or another are surprising or unexpected, there is a strong preference for explanations that rely on single or monolithic causes. Having one rather than several causes for a problem makes it possible to consider each problem by itself and to solve one problem before moving on to the next.

The main assumption of this approach is whatever happens can be decomposed into parts and that each part can be addressed without considering the others. This is so, regardless of whether the problem is one that happened in the past – illustrated by the Root Cause Analysis of an accident or incident – or one that may appear in the future – illustrated by the Fault Tree. In the reactive case, such as accident investigations, the principle is visible in the way that each potential or possible cause is treated by itself, thus leading to (at least) as many solutions or steps to take as there are causes. A good example is provided by an Australian study of how to reduce harm in blood transfusions (VMIA, 2010). The study ended by making 40 different recommendations, distributed as follows: environment (3 recommendations), staff (9 recommendations), equipment (12 recommendations), patient (2 recommendations), procedure (6 recommendations), and culture (8 recommendations), thus nicely illustrating how we try to deal with unwieldy practical problems by simplifying them.

Managing safety by snapshots

An unintended, and usually overlooked, consequence of conventional safety management is that the basis is made up of snapshots of how an organisation functions – or rather, snapshots of how an organisation does **not** function, of how it fails, or has failed. The conventional wisdom is that accidents and incidents provide opportunities to learn and the basis for taking steps to make sure that the same or similar will not happen again. Indeed, one of the seminal works in safety is a book entitled *Learning from Accidents in Industry* (Kletz, 1994). Yet consider for a moment that accidents are events that occur infrequently and irregularly and lead to serious adverse outcomes. Accidents are therefore not typical of how an organisation performs; on the contrary, they represent unusual situations where an organisation has failed either in part or in whole. Yet safety management focuses on and analyses such situations in order to improve safety, following the 'find-and-fix' approach. Without being facetious, it would actually be more appropriate to call it the management of nonsafety.

The principle of safety management based on analysing situations where something has gone wrong can be illustrated by Figure 1.1. The multiple grey traces or curves represent the multitude of ongoing processes or activities that typically take place. The purpose of safety management is to make sure that they do not go below the *limit of safe performance*, shown by a dotted line in Figure 1.1. (Other types of management, e.g., quality management, may try to limit the variability of the processes and keep them as close to a mean value as possible.) The *limit of safe performance* is not fixed, but depends on the current

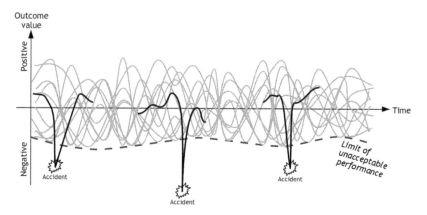

Figure 1.1 Managing safety by means of snapshots.

conditions, and is therefore in Figure 1.1 shown as an undulating curve rather than a horizontal line. The black curve fragments represent processes or developments that produce – or are assumed to produce – the unwanted outcomes. Unwanted outcomes do not occur very often, simply because a functioning organisation cannot afford that. If they occurred frequently, one of two things would happen: steps would be taken to make a change, or the organisation would cease to exist. Neither do unwanted outcomes happen regularly; indeed, when something goes wrong it is usually a surprise; it is unexpected. 'Learning from accidents' is therefore based on haphazard snapshots of situations where the organisation did not work, i.e., on snapshots of irregular anomalies. The fragmented black curves illustrate that we focus on what happened around the accident but not on the long-term developments. To make matters worse, these anomalies are further described in terms of individual 'parts' or structures that have failed, using linear cause-effect relations. It therefore does not seem to be the best basis for managing the safety of an organisation.

Managing safety by everyday work

Instead of basing safety management on the infrequent and irregular occurrences of adverse outcomes and assume that they are the result of orderly 'mechanisms' represented by the fragmented black curves, we should focus on the everyday processes that are represented by the multiple grey curves. These are shown in Figure 1.2, where the graphical rendering has been reversed so that the everyday processes in focus are shown in black while the accidents are shown in grey.

Acceptable outcomes differ crucially from unacceptable or harmful outcomes by being continuous and of course by being acceptable, i.e., above the limit of unacceptable performance. The general purpose of establishing and operating an organisation is indeed to ensure a way of functioning that delivers acceptable

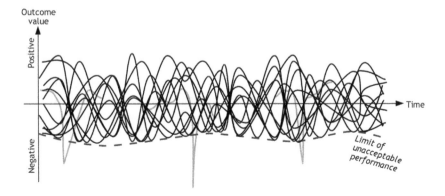

Figure 1.2 Managing safety by means of everyday performance.

outcomes reliably and continuously. The focus of safety management, as well as of management in general, must therefore be the continuous performance or functioning instead of the exceptions, the rare events.

The focus of any kind of management should be on what happens regularly, rather than what happens rarely or not at all. A common argument for the current approach is that the accidents are what happens and that 'nothing happens' when there are no accidents. It is true that nothing spectacular or out of the ordinary happens in periods of calm and stable everyday performance. Nothing *unusual* happens, and nothing happens that automatically attracts attention. But it is a crucial mistake to claim that this means that nothing happens as such. On the contrary, an amazing number of things happen, but they go unnoticed precisely because they happen every day. They escape our attention because they are regular, routine, and habitual and because the outcomes are as expected. But it is precisely when 'nothing' happens that we are safe, in the sense of being without accidents; it is when 'nothing' happens that we are productive, in the sense of generating a sufficient number of output (products) per unit; it is when 'nothing' happens that we are efficient, in the sense of being able to produce with acceptable waste of time and effort; and it is when 'nothing' happens that we are able to produce with high quality, in the sense of having a small number of 'rejects'.

Management of organisational performance in general and safety in particular should therefore be based on an understanding of the 'nothing' that happens all the time, on the typical, everyday processes, rather than on the snapshots of dysfunctional states. Safety management must in particular try to make sure that the typical, everyday processes go well or succeed as often as possible. We need to know and understand how things happen, and we need to be able to measure and assess them. This is what this book is about.

Safety-I and Safety-II

Around turn of the 20th century, ideas about resilience engineering started to spread from a small group of safety experts to the wider community of safety practitioners and academics. At that time, the main motivation was the concern about safety in nontrivial sociotechnical systems with safety defined as the 'freedom from unacceptable harm' or something similar. The development of resilience engineering was driven by the realisation that the established approaches to safety were ineffective and possibly even harmful and therefore a hindrance for progress (Haavik et al., 2016). Safety management was based on a strong belief that all adverse outcomes had identifiable causes and that these causes could be eliminated or neutralised once they had been found, as explained by the *causality credo*. Safety was defined generically as a condition where as few things as possible went wrong. It followed from this understanding that safety could be achieved by *preventing* things from going wrong, i.e., by looking at and responding to unwanted outcomes and the events that were assumed to lead to them. This understanding of safety is now called Safety-I (Hollnagel, 2014a).

Resilience engineering took a different approach. It started by disputing the tacit assumption that there was a value symmetry between effects and causes, in the sense that adverse outcomes were due to similarly adverse causes. Value symmetry expresses the belief that unacceptable outcomes have unacceptable causes and vice versa. An accident is thus seen as a consequence of something having gone wrong – a failure, a malfunction, or an error. It corresponds to our moral codex that good deeds are rewarded while bad deeds are punished. In this case, the 'rewards' and 'punishments' are the acceptable and unacceptable outcomes, respectively, while the deeds are the inferred causes. This view is also called the hypothesis of different causes, meaning that the causes of acceptable outcomes are different from the causes of unacceptable outcomes. Unacceptable outcomes are caused by failures, malfunctions, or errors while acceptable outcomes are 'caused' by correctly performing systems, in particular by correct human performance. Instead, resilience engineering argued that 'failures were the flip side of successes', or in other words that things that go right and things that go wrong happen in basically the same way. Consequently, the implicit hypothesis about value symmetry had to be abandoned.

By abandoning the assumptions of value symmetry – which can be seen as the moral equivalent to the hypothesis of different causes – there is no longer any need to invoke 'error mechanisms' that appear *deus ex machine* when something goes wrong but otherwise lie dormant. Efforts to try to prevent something from going wrong should therefore be replaced by efforts to ensure as far as possible that everything goes well. This leads to the definition of Safety-II as the situations where as much as possible goes well. In this interpretation, safety is about how to support, augment, and facilitate the everyday activities that are necessary for acceptable outcomes on all levels of an organisation. It is about a kind of safety that is inseparable from quality and productivity, which means that these interests require measures and methods that are compatible rather than incompatible.

Safety as an antidote

The English word 'safe' comes from the French word *sauf*, which means without, which in turn comes from the Latin word *salvus*, which means intact or whole. This is the foundation for the current meaning of safe as being without harm or injury. (The American National Standards Institute, for instance, defines safety as the freedom from unacceptable risk.) Safety has therefore been prescribed as an antidote to the causes of accidents, incidents, and adverse outcomes in general. The antidote has necessarily been relative to the commonly agreed causes of harm and injury and the ways in which they could exert their roles. This can easily be illustrated by looking at the major trends in safety management.

Antidote #1: prevention and elimination

The first trend corresponds to the simple linear thinking represented by the Domino model. This safety paradigm proposes that there is a recognisable cause-effect relationship between causes and effects, in some cases as a simple cause directly leading to an effect, but in most cases via a chain or sequence of cause-effect relations. According to this way of thinking, safety can be achieved by preventing things from going wrong, by preventing causes from exerting their effect, by preventing hazards from becoming realised, or by eliminating causes and hazards. The first thing to do is therefore to identify the possible hazards or sources of risk, and then to determine or decide whether they are large enough to require attention. This is still the most common basis for the management of safety today and has been enshrined in such standard methods as Failure Mode and Effects Analysis, Human Reliability Assessment, and Hazard and Operability Study.

Antidote #2: improved defences

The second trend corresponds to composite linear thinking as represented by the Swiss cheese model. This safety paradigm proposes that adverse outcomes are due to a combination of adverse events and a failure of precautions, barriers, or defences. The solution is therefore to strengthen the barriers and defences, or to have multiple defences or defence-in-depth. As long as the causes involve physical movement or physical acts of some kind, physical and functional defences can be used with acceptable efficiency. But when the causes refer to decisions, choices, priorities, etc., or even to an organisations' culture, the use of defences becomes a bit problematic since the barriers must be symbolic or incorporeal rather than physical and functional.

If it sometimes is difficult to make the defences concrete, it is even more difficult to make some of the causes concrete. An example is the concept of drift, as in the idea of 'the normalisation of deviance'. But an organisation cannot 'drift', one reason being that there is no well-defined entity in the first place, and another that there is no reasonable definition of the 'space' in which it can drift. This does not deny that there may be slow and gradual changes in working practices, which may affect how an organisation performs and which in hindsight may be interpreted as an erosion of safety buffers. But these changes can better be understood as gradual shift in the trade-off between efficiency and thoroughness to the advantage of the former rather than the latter.

Antidote #3: coping with complexity

The third trend corresponds to the contemporary fascination with complexity and complex systems. This has developed gradually since the beginning of the 1980s, when Charles Perrow (1984) introduced the concept of normal accidents and pointed out that some systems were complex because they were characterised by tight couplings and nonlinear relations. But if we accept that complexity is a genuine feature of some contemporary systems – most of which are large-scale sociotechnical systems – the question of the antidote remains. Unlike the first and second trends, the third trend – complexity – does not point to any recognisable causes. Outcomes are not longer the result of identifiable causes but rather emerge from complexity itself. But what are the solutions? Outwitting complexity? Reducing complexity to make systems simpler? Augmenting the feeble human mind with automation?

To understand what an antidote might be, it is necessary to recognise that there are several different types of complexity:

- Mathematical complexity, which is a measure of the number of possible states a system can take on, when there are too many elements and relationships to be understood in simple analytic or logical ways.
- Pragmatic complexity, which means that a description, or a system, has many variables.
- Dynamic complexity, which refers to situations where cause and effect are subtle and where the effects over time of interventions are not obvious.
- Ontological complexity, which has no scientifically discoverable meaning, as it is impossible to refer to the complexity of a system independently of how it is described.
- Epistemological complexity, which can be defined as the number of parameters needed to describe a system fully in space and time. While epistemological aspects can be decomposed and interpreted recursively, ontological aspects cannot.

The relation between ontological and epistemological complexity is fundamental to understanding what complexity is (Pringle, 1951). If we assume that there is a true ontological complexity, which means that some systems (or phenomena) by nature are complex, then we need to consider whether it is possible to give a simple description of that complexity. If the answer is yes, then it stands to reason that complexity, as a quality or a phenomenon, is independent of its description. In that case, we should be able to understand complexity as it is and not as it is described. But if the answer is no, then we must accept that complexity is a quality of how something is described, hence epistemological rather than ontological. There is therefore no logical basis for assuming that there is a quality of complexity that is independent of and separate from its description. A 'complex system' is therefore nothing more than 'a system with a complex description' or a system that is intractable rather than tractable. In that case, there is obviously no need of an antidote to complexity, since complexity is nothing more than a complicated description.

The distinction between complex and complicated has practical importance. From a practical point of view, a system can be called complicated if the description requires many terms or parameters and if changes in one may affect many or all others significantly. In the same way, a system may be called complex if some of the parameters are unknown and/or unknowable so that it is in effect impossible to measure or control them. For a complex system, the solution is to try to make sure that all the parameters are known and to de-couple parameters when it is possible. If this can be achieved, the system ceases to be complex (since it is an epistemic category in any case) and instead becomes 'merely' complicated or nontrivial.

While the transformation from a complex to a complicated system may be possible for technological systems (cars, oil refineries, train networks, etc.), it is not possible to use this solution for sociotechnical systems or for organisations. In any large-scale social system, there will always be couplings that are un-known. A simple reason is that such systems are not designed but develop or even grow from an initial basis. The growth is however partly autonomous (and teleological), hence impossible to describe in detail.

The real antidote to complexity lies in realising that it is epistemological rather than ontological. Complexity is used as a label to disguise our limited ability to understand and describe the consequence of the changes that we make, which shape the conditions in which we live and act. But if we recognise complexity for what it is – a euphemism that we use to beautify our ignorance and cognitive limitations – the solution becomes straightforward. The solution is not to counter the ontological complexity of the real world, since there is none, but rather to improve the means we have to describe the world. This description is currently dominated by linear cause-effect thinking. Complexity may, indeed, be seen as an artefact of linear descriptions. We therefore need to go beyond that and develop concepts and relations, a 'language' that enable us to represent the unpredictability and uncertainty that limits current ap-proaches. That the human mind is able to do so has been demonstrated rather conclusively by mathematics over thousands of years.

Chapter 2

What does 'resilience' mean?

The aim of this book is to present a practical way to develop and manage an organisation's resilience potentials in order to ensure – as far as possible – that the organisation can cope with the expected and unexpected conditions it is certain to encounter. The writings about resilience engineering have from the very beginning made clear that resilience was associated with what an organisation *did* – how it performed – and not with what that an organisation *had* – what it was. Resilience refers to a characteristic way in which an organisation performs – meaning *what* it does and *how* it does – rather than to a characteristic of an organisation *per se*, to a quality or to something that an organisation *has*. In spite of that, much energy has been used – or wasted – to discuss what resilience is, how resilience can be measured, and how resilience can be engineered or managed.

The way in which an organisation performs clearly depends upon what it is capable of doing. Actual performance depends on potential performance, and the former can reasonably be interpreted as a subset of the latter. Resilient performance can be seen as the synthesis in an actual situation or condition of an organisation's coping potentials; resilient performance is unlikely to be the expression of a single factor or ability but rather represents a nontrivial combination of several factors or potentials. The question in practice, and therefore for this book, is how it is possible to develop and manage these potentials.

The origin of resilience as a concept

It is widely accepted that the term resilience was first used by the British Navy in the beginning of the 19th century as a way of explaining why some types of wood were able to accommodate sudden and severe loads without breaking (Tredgold, 1818). Resilience therefore represented a characteristic or quality of materials. About 150 years later, the Canadian ecologist Crawford Holling (1973) proposed that ecological systems could be described in terms of two distinct properties called resilience and stability, respectively. In ecology, the term *resilience* referred to a system's ability to absorb changes, while the term *stability* referred to a system's capacity to return to an equilibrium state after a temporary disturbance. The definition of ecological resilience was later expanded through the introduction of the notion of the adaptive cycle (Carpenter et al., 2001). This led to the conclusion that (ecological) resilience had three primary properties:

- The amount of change a system can undergo while maintaining the ability to function.
- The degree to which a system can organise itself.
- The degree to which a system can develop a capacity to learn and adapt.

Adaptation and the concept of an adaptive cycle continue to play a prominent role in contemporary discussions of system theory, for instance, in the guise of Complex Adaptive Systems, and have been much discussed in resilience engineering.

It makes sense to argue that ecological systems need an adaptive cycle since they evolve in a changing environment. (An important proviso is, of course, that the rate of adaptation must be markedly faster than the rate of change in the environment.) But an adaptive cycle is not necessary for the sociotechnical systems that constitute the focus for resilience engineering. The reason is simply that since ecological systems are not conscient and therefore unable to have intentions, they are limited to being reactive. Sociotechnical systems, organisations, on the other hand, are conscient because they – by definition – include humans. Compared to ecological systems, sociotechnical systems can and do consider what may happen and can and do use that to direct their behaviour. They are therefore less dependent on repeated adaptations or an adaptive cycle. (Needless to say, they may not all perform proactively equally well, and some manifestly do it rather badly.) Anticipation is faster and far more powerful than adaptation and can therefore literally short-circuit the adaptive cycle. The disadvantage is, of course, that the anticipation refers to the future and therefore is not completely certain. This risk can, however, be outweighed by improving the potential to anticipate by learning from past examples and by limiting the time horizon.

Other uses of resilience

Outside ecology, the term 'resilience' began to be used in the early 1970s as a synonym for stress resistance in psychological studies of children and to describe the human capacity to withstand traumatic situations in general. Towards the end of the century, it was picked up by the business community to describe the ability dynamically to reinvent business models and strategies as circumstances change. Today references to resilience can be found in economics, pedagogics, psychology, sociology, risk management and network theory – and possibly in other places as well.

The transition from ecology to business, in particular, changed resilience from being a passive to becoming an active characteristic. In the business context, resilience describes the capacity to survive, to adapt, to grow in the face of turbulent changes and to change before change becomes inevitable (Hamel and Välikangas, 2003). Because a business environment can change rapidly, resilience cannot rely passively on adaptation or adaptive cycles but requires active management and feedforward control based on anticipated outcomes. This highlights the crucial difference between non-intentional (ecological) and intentional (sociotechnical) systems. The former are limited to responding to whatever happens while the latter can act in anticipation of what may happen – before the need becomes desperately obvious. The former are reactive while the latter can be both reactive and proactive. In the former, resilience is a natural trait; in the latter, it is an engineered or deliberate trait.

Negative connotations

The common use of the term resilience, within safety and elsewhere, has a negative connotation in the sense that it focuses on how an organisation handles diversity, stresses and disruptions. The negative connotations are understandable considering that the origin of resilience, or the first modern use of the term, was in physics and material science. Since a physical material is passive, since it can only respond, resilience must be seen in relation to, or as a reaction to, the potentially harmful or destructive consequences of unexpected events. Even when the idea of resilience was picked up by ecology, the focus on adversity remained. Although an ecological system is dynamic where a physical system is static, it is still not conscient. It remains reactive, perhaps with an element of random responses thrown in, and only acts in response to something that happens, to something that in one way or another is imposed by external forces or agents.

When resilience was introduced to the field of industrial safety (in a wide sense), as in resilience engineering, the negative and reactive connotations were carried along. That is not hard to understand either, since safety – or rather Safety-I – as described in Chapter 1, traditionally has been preoccupied with avoiding adversity, risks and harm. The focus on the negative, on sustaining an organisation despite adversity, is very common, and resilience is therefore seen as the ability of an organisation to react to and recover from disruptions with minimal effect on its dynamic stability.

Resilience engineering, however, is not only about dynamic systems but also about sociotechnical systems. It is about organisations as deliberate or intentional configurations of people, materials and activities (and information) that serve to achieve a given objective or objectives. But this means that it is no longer sufficient to think of resilience vis-a-vis the negative. In order for an organisation to exist and to continue to exist, it must not only respond *when* something happens but also act *before* something happens. And it must not only act in the face of dangers, trying to protect itself, but also in the face of opportunities that allow it to survive and grow – in every possible way. Recognising and responding to opportunities is an indispensable part of everyday activity throughout an organisation for individuals, for social groups (collectives), for management and for an organisation itself. In that sense, there is a similarity to ecological systems, which also grow when the opportunity arises. But the difference is that ecological systems neither look for nor anticipate opportunities. Most importantly, they do not try to create or bring about opportunities, but organisations do. Organisations are – or can be – strategic and tactical.

Accentuate the positive

As the above brief history shows, the thinking about resilience has typically referred to a dichotomy: on the one hand, materials, systems or situations where resilience was absent and where adverse outcomes therefore might result, and on the other hand, materials, systems or situations where resilience was present and where adverse outcomes could be avoided. This was also the case in the early 2000s when resilience engineering was proposed as an alternative (or rather as a complement) to the conventional view of safety. The first book, *Resilience Engineering: Concepts and Precepts*, provided the following definition.

> The essence of resilience is therefore the intrinsic ability of an organisation (system) to maintain or regain a dynamically stable state, which allows it to continue operations after a major mishap and/or in the presence of a continuous stress.
>
> (Hollnagel, 2006)

The definition reflected the historical context by its juxtaposition of two states – one of stable functioning and one where the system had broken down. But this also limited the definition to situations of threat, risk or stress.

Discussions about resilience versus robustness or resilience versus brittleness nevertheless soon made clear that resilience is not just about avoiding failures and breakdowns or the converse of a lack of safety. In a later book, *Resilience Engineering in Practice*, the definition changed to this:

> The intrinsic ability of a system to adjust its functioning prior to, during, or following changes and disturbances, so that it can sustain required operations under both expected and unexpected conditions.
>
> (Hollnagel, 2011)

In this definition the reference to risks and threats was replaced by a reference to 'expected and unexpected conditions'. The focus had also changed from 'maintaining or regaining a dynamically stable state' to the ability to 'sustain required operations'. The logical continuation of these developments leads to the following definition:

> Resilience is an expression of how people, alone or together, cope with everyday situations – large and small – by adjusting their performance to

the conditions. An organisation's performance is resilient if it can function as required under expected and unexpected conditions alike (changes/disturbances/opportunities).

The changes in the definitions broaden the scope of resilient performance. It is not just about being able to recover from threats and stresses, but rather about being able to perform as needed under a variety of conditions – and to respond appropriately to both disturbances and opportunities. The focus of resilience engineering is thus resilient performance, rather than resilience as a property (or quality) or resilience in an 'X versus Y' dichotomy.

The emphasis on opportunities is important for the change from protective safety (Safety-I) to productive safety (Safety-II) – and ultimately for the dissociation of resilience from safety, thereby leaving the sterile discussions and stereotypes of the past behind. Resilience is about how organisations perform, not just about how they remain safe. An organisation that is unable to make use of opportunities is not in a much better position than an organisation that cannot respond to threats and disturbances – at least not in the long term.

The purpose of resilience engineering is to ensure that an organisation can perform effectively in everyday conditions, in other words do everyday work successfully. It is also about ensuring that an organisation is able to cope with more unusual situations – the unexpected ones – both when they have the potential to disrupt functioning or performance (threats and risks) and when they have the potential to improve or augment everyday performance (opportunities). Indeed, resilience is found not only during non-routine or critical incidents but also, and perhaps more so, when incidents are not happening, i.e., during routine functioning with acceptable outcomes. Ecology and business agree that the resilience of an organisation is an expression of its ability to sustain its own existence – to survive and to thrive. The survival or continued existence is in itself proof of the resilience but not synonymous with it.

How resilient is my organisation?

As soon as it became accepted that an organisation's resilience was an important issue, questions were asked about how the level or degree of resilience could be determined. The analogy with safety culture and other monolithic concepts was irresistible. And since it was common practice to refer to the levels of safety culture, it was taken for granted that a similar question could be asked about the levels or degrees of resilience. But just as it should be asked whether the concept of levels of safety culture is meaningful, it should be asked whether the concept of levels of resilience is meaningful. In neither case does the concept of distinct levels – of safety culture or resilience – make much sense. In the case of safety culture, it is probably too late to change this widespread belief, but in the case of resilience it is (hopefully) not.

Rather than talking about resilience, viz. the arguments above that resilience as a quality does not exist, we should talk about an organisation's potentials for resilient performance or – slightly incorrectly – an organisation's resilience potentials. Whereas an organisation cannot have resilience, it can have the potentials for resilience, or more precisely for performing in a manner that can be characterised as resilient. This corresponds well to the updated definition of resilience discussed above.

An organisation's potentials for resilient performance do not necessarily mean that it always will perform in a resilient manner. But the reverse relation holds. Performance cannot *consistently* be resilient unless an organisation has the potentials for resilient performance. In other words, while establishing and managing the potentials does not by itself guarantee resilient performance, the lack of the potentials will make resilient performance very unlikely – unique cases excluded.

With this preamble, the rest of the book will be about the potentials for resilient performance – how it is defined, how it can be measured and managed and how this provides a proven and viable alternative to today's safety managements systems.

Chapter 3

The basis for resilient performance

An organisation can be defined as a stable association of people (called members of the organisation) who engage in concerted activities in order to attain specific, common objectives. An organisation is managed by assigning different roles or functions to different groupings or sets of members and coordinating their work so that the goals are achieved in a way that fulfils one or more criteria. These criteria can be material with regard to safety, productivity, quality, speed or precision or personal with regard to well-being, a sense of achievement, development, etc. To sustain its performance, an organisation – or rather the management of an organisation – must be able to coordinate activities, ensure that the resources for effective work are available and resolve possible conflicting goals and priorities between the people and groups.

An organisation is heterogeneous rather than homogeneous because people either are assigned to or assume different roles and responsibilities. This means, among other things, that one part of an organisation has the responsibility of managing what other parts of an organisation do – and in some cases also of managing itself. The primary purpose of management is to coordinate the efforts of an organisation, or strictly speaking of the people in an organisation either including or excluding the management itself, so that the goals and objectives are accomplished efficiently and effectively.

Work-as-Imagined and Work-as-Done

For anyone who is responsible for managing others, a central question is what determines what other people do. This is important when *planning* what people should do in order for work to be successful (to produce the intended outcomes), when *managing* what people do while carrying out the work taking into account the actual conditions of work and when *analysing* the outcome, specifically the situations where the outcome was unacceptable and/or unexpected. The assumptions or expectations of what other people should do is called Work-as-Imagined (WAI), while that which people actually do is called Work-as-Done (WAD). In the planning of work, in the management of work and in the analysis of work, it is always necessary to refer to a description of work as it is assumed to be – as it is imagined. This situation is shown in Figure 3.1.

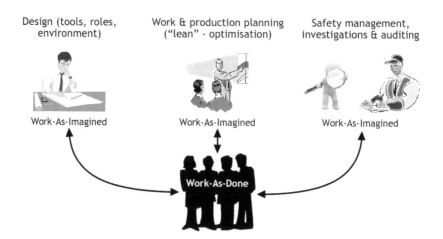

Figure 3.1 Work-as-Imagined and Work-as-Done.

The term 'imagined' is not used in a uncomplimentary or negative sense but simply recognises that our descriptions of work will never completely correspond to work as it takes place in practice – as it is actually done. Even when significant efforts are made to standardise work and working conditions in order to make work as regular and predictable as possible, there will always be a number of differences – most of them small but some of them large. Since the actual management of work obviously must refer to what in fact is happening, and not to what we assume is happening (or what we hope will be happening or believe has happened), it is essential that the management of work, and *a fortiori* the management of an organisation, is based on a description or representation – a model – that is as realistic and accurate as possible.

A condition for effective management is therefore that managers can correctly predict how the people they manage will behave. This goes regardless of whether the relationship is one-to-one as in a master-apprentice relation or many-to-many as in the management of a small group of people, of a department or unit, of a division, of a company or of a multinational enterprise. And it goes regardless of whether the purpose or aim is safety (reduction of harm and injury as in Safety-I), timely delivery (reduction of delays), efficiency (reduction of waste), quality (reduction of imprecision or slack), productivity (reduction of downtime, reduction of cost per unit produced), customer satisfaction or for that matter anything else. In order for such predictions to be better than random guesses, it is necessary to have a model or a systematic – perhaps even scientific – description of why people do what they do.

Individual or organisational 'mechanisms'?

Philosophers and economists through the ages, more recently joined by behavioural and organisational scientists, have recognised the necessity of understanding what it is that makes people do what they do. The answers have been many and varied and have inevitably reflected the dominant thinking of the time. Leaving philosophy and economy aside – and also limiting the discussion to the last century or so – the debate was initially one of the 'nature of man', of individual motivation or 'mechanisms', of what makes people 'tick', but was later expanded to include a discussion of the influences from the surroundings and in particular from the organisation.

A famous, and often reviled, example of that discussion is found in *The Principles of the Scientific Management* developed by F.W. Taylor in the first decade of the 20th century (Taylor, 1911). The starting point was the problem of how best to increase the productivity of people at work, for instance, brick layers, iron yard workers and manual inspection. Taylor argued that it was better to optimise the way that work was done than just to make people work as hard as possible. Scientific Management thus in a way introduced the interest for WAD, although the purpose was to use this to prescribe an optimal way of working (WAI) and then change WAD to achieve that. The steps to change WAD were simple:

- Determine the most efficient way to perform specific tasks, match workers to their jobs based on capability and motivation and train people to work at maximum efficiency.
- Constantly monitor performance to ensure that people use the most efficient ways of working and to avoid soldiering (loafing). This implies a clear differentiation and allocation of work between managers and workers. Indeed, Scientific Management required a higher manager-to-worker ratio than previous management methods.

Scientific Management took for granted that people were motivated by money, leading to the idea of 'a fair day's pay for a fair day's work' that still is with us. Scientific Management represents an engineering perspective on work and people and can be seen as heralding the ideas about reduction of waste that became popular 20 years later and have remained popular ever since. This is clear from the preface to the book that provided the foundation for quality control, where Walter Shewhart (1931) declared, 'The object of industry is to set up economic ways of satisfying human wants and in so doing to reduce everything possible to routines requiring a minimum amount of human effort'. Despite this enthusiasm, the engineering perspective did not go unchallenged at the time and is still a contentious issue. The alternative is, of course, a humanistic approach where the starting point is *people* rather than *work*. The premier example of that is the theory of human motivation (Maslow, 1943). The central idea was that people have a need to actualise themselves as fully as possible; the focus was therefore well-functioning individuals or even exemplary people rather than people engaged in manual work. But Maslow's ideas also influenced the understanding of worker behaviour, and his ideas about self-actualisation even became the basis for a specific theory of *eupsychian* management (Maslow, 1965).

Maslow posited that people's behaviour was motivated or driven by their needs, ranging from fundamental needs to the need for self-actualisation. The needs were described as forming a hierarchy, usually represented as a pyramid with physiological needs at the bottom and the need for self-actualisation at the top. In between were the needs of safety, the needs of love and belonging and the needs of self-esteem. According to this theory, the basic needs must be met before a person will pursue the needs higher in the hierarchy. Pursuing higher-level needs, however, requires that lower-level needs continue to be fulfilled, which means that the person must be able to satisfy multiple needs at the same time. The implications for the management of work were to ensure that the lower-level needs were fulfilled in order for the higher level needs to be addressed. Unlike Taylor, people are not primarily seen as seeking monetary rewards, although 'a fair day's pay for a fair day's work' still is necessary for people to become engaged in their work. What makes people 'tick' is the need for self-actualisation, and work management must ensure that this is possible.

A third approach, which juxtaposed the engineering thinking of Taylor and the humanistic thinking of Maslow, was Douglas McGregor's (1960) proposal of two theories, 'Theory X' and 'Theory Y', that represented contrasting models of workforce motivation. According to 'Theory X', people are naturally unmotivated and dislike working. This encourages an authoritarian style of management that actively intervenes to get things done. Management must trace all actions and reward or reprimand the responsible individual according to the actions' outcomes. A flaw of this management style is that it limits the employee potential and discourages creative thinking. The alternative is 'Theory Y', which argues that people enjoy their physical and mental work and have the ability to solve problems in a creative way. According to Theory Y, their talents are wasted if management is based on standards, rules and restrictions. Instead, managers should promote the optimum workplace through morality, creativity, spontaneity, problem-solving, lack (or minimisation) of the effects of prejudice and acceptance of facts.

The intention of 'Theory X' and 'Theory Y' was to characterise two sets of beliefs that a manager might hold about the origins of human behaviour, but not to suggest that one was superior to the other. 'Theory Y' pointed forward to the current understanding of work in a sociotechnical system. But traces of 'Theory X' can still be found, for instance, in the belief in a zero accident culture and indeed in the Safety-I perspective.

From individual performance to culture

Understanding human performance, however, requires more than a theory of the individual. What people do depends also on the influences from the surroundings and on demands, expectations, norms and values – in other words on the organisation. Notwithstanding the many theories about the nature of humans, ranging from the human as a stimulus-response mechanism (behaviourism) to the human as an information processor (human information processing), the fact remains that what people do depends on their social and organisational environment as much as on how they think and feel. People do not act on what they can see, on what is actually there and on what they have been taught. They act on what they perceive, on what they pay attention to and on what they can remember. But what they perceive, what they attend to and what they remember are based on multiple and sometimes conflicting interest and motivations and are rarely if ever in agreement with the ideal of rational decisions. What people do reflects their understanding of the situation, their socially conditioned assumptions about how the 'world' works (model, causality), their time horizon, their interests – individual and altruistic and social, situational pressures, as well as many other things.

With the usual preference for explanations that refer to a single cause (cf., the section on monolithic explanations later in this chapter) and simplified deductions, this has been described as the culture or the organisational culture. One of its early expressions was the notion of an *Esprit de Corps*, the idea that the soldiers in an army shared a strong team spirit, a sense of duty and a devotion to a cause. The *Esprit de Corps* was an important trait of the *Légère* regiments in Napoleon's *Grande Armée*. But something similar can also be found in the military culture of the Roman legions and even earlier in the Spartan army. In psychological terms, the *Esprit de Corps* can be seen as an expression of a shared level of aspiration (*anspruchsniveau*) or the norm for acceptable performance (Chapman and Volkman, 1939). While the level of aspiration usually is seen as an individual characteristic, it obviously depends on what the person assumes that the people around expect or what an organisation requires.

The dependence between the beliefs of an individual and the beliefs of the organisational environment also plays a prominent role in a critical anthropological analysis of theories of culture by Keesing (1974). Cultures are defined as systems of '(socially transmitted behavior) patterns that serve to relate human communities to their ecological settings' (p. 75). Cultures, in other words, represent the common agreement on how to behave and what to do both generally and in specific conditions. These 'behavior patterns' are based on an individual's 'theory of what his fellows know, believe, and mean, his theory of the code being followed...' (p. 89). An organisation was seen as the carrier of distinct cultures, or sets of shared values, beliefs and norms that guided the attitudes and actions of its members.

The anthropological thinking about cultures, as well as the early social-psychological thinking about cultures, focused on the value or usefulness of culture but did not treat culture as a subject in itself. The use of the concept of organisational culture started in management and organisation studies in the late 1970s and became common in the 1980s. This received a significant boost when the concept of a safety culture was proposed in response to the accident at Chernobyl in 1986. In the study of High Reliability Organisations (HRO), it was noted that culture 'creates a homogeneous set of assumptions and decision premises which, when they are invoked on a local and decentralized basis, preserve coordination and centralization. Most important, when centralization occurs via decision premises and assumptions, compliance occurs without surveillance' (Weick, 1987).

The present-day concept of organisational culture reached its final form when the psychologist Edgar Schein defined it as '... (a) a pattern of shared basic assumptions, (b) invented, discovered, or developed by a given group, (c) as it learns to cope with its problems of external adaptation and internal integration, (d) that has worked well enough to be considered valid and, therefore (e) is to be taught to new members as the correct way to perceive, think and feel in relation to those problems' (1990). Schein's model proposed three distinct levels in organisational cultures characterised by how easily they were visible to an observer. The easiest to see are *artefacts*, which include the tangible, overt or verbally identifiable elements in an organisation, e.g., architecture, furniture, dress code, etc. Artefacts can be recognised even by people who are not part of the culture. Next are the *espoused values* such as the identity, official targets, policies and rules of behaviour. (The most controversial example is probably 'safety first'.) An organisation's members use the *espoused values* to represent the organisation to themselves and to others and therefore also as an expression of what they – and others – hope to become. The final level is made up of the *shared basic assumptions*, that which is taken for granted, which usually are unconscious (or incorporeal) and which in a sense constitute the essence of culture. The shared basic assumptions are often easier to notice from without than from within the organisation.

It is clear that the practical work with cultures must comprise individual attitudes towards safety as well as the shared attitudes and the shared understanding, including the organisational structures and resources that create and support these attitudes. The thorny question here is whether a change to the culture will lead to a change in people's performance, or whether it is the other way around. If we think that the main determinant of what people do is the safety culture – or the organisational culture – then culture is the independent variable and performance the dependent. But we might also consider, whether it is the other way around, that the culture is affected by what people do and that the culture mainly is a synthesis or abstraction of performance. This discussion will be resumed in Chapter 7.

Culture as a suffix

The term 'culture' is used and misused in many ways. It often serves as a convenient solution to name something that is assumed to be important for an organisation's performance, usually but not exclusively safety, but which is not completely understood. This can be seen from the many ways in which 'culture' is used as a suffix, for instance, to define the various subspecies of safety culture.

- Safety culture – the ways in which safety is managed in a workplace, often described as *'the attitudes, beliefs, perceptions and values that employees share in relation to safety'* or simply 'the way we do safety around here'. This, of course, begs the question of what kind of safety it is.
- Reporting culture – the willingness to produce, collect and analyse reports about accidents and incidents, about things that can be a risk for an organisation.
- Just culture – the principle that front-line operators and others are not punished for actions, omissions or decisions taken by them that are commensurate with their experience and training, but where gross negligence, wilful violations and destructive acts are not tolerated.
- Learning culture – the organisational conventions, values, practices and processes that serve to encourage employees and organizations to develop knowledge and competence.
- Security culture – the shared customs of a community that serve to minimise the risks that community activities are subverted or targeted for sabotage.

These, and possibly other, examples illustrate that while it is simple to propose an X culture as an *explanandum*, it only serves to give a name to something but does not explain it. One way to compensate for this slightly embarrassing fact is to present a measurement of X culture, since a measurement 'obviously' proves that whatever is measured really exists. Humans have a seemingly insatiable need to measure something in order to prove that they understand it and that they can control it. This is a reflection of Lord Kelvin's famous dictum that 'To measure is to know'. Measurements certainly help to reduce the dreaded lack of comfort that follows the unknown, and not knowing what safety culture is (since it has become the *sine qua non* in safety management), is uncomfortable. But measurements of safety culture – and even worse, of levels of safety culture – do not make safety culture real, nor do they prove its existence.

The preference for simple and single solutions, such as safety culture, is clearly wrong and misses the target. Organisational culture, for lack of a better term, is undoubtedly an important determiner of performance, but it is not the only one. Indeed, as Schein's theory points out, organisational culture has several aspects, some visible (and therefore easy to change) and some less visible, and therefore more difficult – or even impossible – to change.

The resilience potentials

It is obvious that the performance of an organisation in some fundamental sense is inseparable from the performance of the people that *are* the organisation. This goes for every aspect of performance and therefore also for resilient performance. But it is also obvious that multiple factors determine the performance of an organisation and the people in it. It is not just external conditions (simple and nonnegotiable standards, time-and-movement studies), nor is it just internal or psychological factors, whether self-actualisation or something else, nor for that matter cognitive factors, striving for rationality or simple hedonism.

Resilience engineering has from the very beginning adopted a functional perspective, looking at what an organisation *does* rather than at what it *is*. Whereas safety management and safety culture, and to some extent also the school of HRO, justify their existence by referring to things that go wrong, or in other words a Safety-I perspective, resilience engineering focuses on how an organisation performs and on whether that performance contributes to the sustained existence of the system by enabling it to carry out its functions under expected and unexpected conditions alike. An organisation can survive in several ways: by sustaining or upholding present functioning, by growing or extending the present functioning (such as market growth) or by changing and developing so that it can function in new ways – like companies that maintain the name and the brand but take up entirely new areas of business. Resilience engineering looks at everything an organisation does, at its functioning on a broad scale (of outcomes) and not only at what goes wrong. Safety is neither the only nor the primary concern, but just one among several.

In the list of X cultures above, there was no entry for a resilience culture (or a culture of resilience) although the term is used with increasing frequency. While it clearly is important to understand why an organisation can perform in a manner that is resilient, the solution is not to look for or to postulate the existence of a resilience culture. To find the answer, it is necessary to look more closely at what enables an organisation to perform in a certain manner. Resilience engineering proposes that four potentials are of primary importance: the potential to respond, the potential to monitor, the potential to learn and the potential to anticipate. The only way in which a label such as 'resilience culture' can make sense is therefore as a reference to the way in which these potentials contribute to the daily practice. However, the label is better avoided altogether.

Interlude: on monolithic explanations

The preference for single and simple explanations is ubiquitous in how humans strive to account for what happens around them, not least when well-intentioned actions lead to unexpected consequences. It can be found in all fields of activity, politics, ethics, law, biology, history, finance, science and of course in industrial safety. In the latter case, it is convincingly illustrated by the categorisation of safety thinking as having three ages called the age of technology, the age of human factors and the age of safety management, respectively (Hale and Hovden, 1998). This categorisation highlights the fact that in each age a single explanation or cause – technical failures, 'human error' and safety culture, respectively – was accepted as the predominant solution to a host of problems.

Such many-to-one solutions are obviously attractive since it becomes easier both to explain what has happened and to communicate it to others. While the practical value in many cases is limited, the emotional value, the ability to set the mind at ease, is indisputable. Explanations of this nature can be called monolithic because they rely on a single concept or factor. It is not really surprising that monolithic thinking is so pervasive, since it corresponds to the very language that is used to describe how humans think. We do talk about a (single) line of thought or a (single) line of reasoning. Indeed, the very ideal of reasoning is logical thinking, which is strictly linear.

Monolithic explanations can be seen as representing a social convention and are therefore essentially social constructs. They can also be seen as a form of an Efficiency-Thoroughness Trade-Off (Hollnagel, 2009a). Monolithic explanations are efficient to use, they are quick to apply and require little cognitive or mental effort, but they lack in thoroughness and in precision. This will, sooner or later, show itself as an inability actually to improve the situations, which the monolithic explanations seemed to resolve. Monolithic explanations are, of course, also the ultimate solution to input information overload problems (Miller, 1960) – it reduces the categories of discrimination to only one.

The common approaches to safety and safety management abound with monolithic explanations. In addition to 'human error' and safety culture, frequently used exemplars are situation awareness, complex adaptive systems and unfortunately also resilience. They all share the quality of being intuitively meaningful, and their face validity is therefore seldom questioned. They also resemble articulated scientific concepts because they are treated in a voluminous scientific literature and explained by impressively named nonobservable theoretical constructs. But in fact they are just 'folk models' or safety myths (Dekker and Hollnagel, 2004; Besnard and Hollnagel, 2012), something that is take for granted but is not – and cannot be – verified. Monolithic explanations are generally used on their own to provide a single solution to a problem. The solution can be that something was missing or absent – a lack of situation awareness or a deficient safety culture – or conversely that something was present – that people made a 'human error' or that the situation or the system was complex. In either case, the simple solution leads to a simple response – either provide what was missing or eliminate what was present.

Chapter 4

The resilience potentials

The purpose of managing work is to ensure that acceptable outcomes occur with the intended frequency, speed and reliability – which also means that the number of unacceptable outcomes is kept to a practical minimum, if not completely prevented. This can only be done if there is a reasonably correct understanding of how an organisation functions and how people perform at work – of what determines Work-as-Done. The understanding is the necessary basis for efforts to make some outcomes – the desirable ones – more likely and others – the undesirable ones – less likely.

As argued in Chapter 3, human and organisational performance depends on many different things that cannot reasonably be collapsed into a single factor. Resilience is a characterisation of certain kinds of performance, and the most one can say is that an organisation may have the potentials for resilient performance. According to the current definition of resilience, this requires an ability to adjust performance to the conditions, an ability to respond to changes, disturbances and opportunities, and an ability to do so in a flexible and timely manner. Further, it is required that these abilities are in play prior to, during and after something happens. On the basis of these definitions, it is possible to propose some potentials that are necessary and sufficient for an organisation to perform in a resilient manner. Since resilient performance is possible for most, if not all, organisations, the potentials must be independent of any specific domain. (Resilient performance is, of course, something that also applies to individuals, although this is not the main focus here.) Resilience engineering has proposed that the following four potentials are necessary for resilient performance (Hollnagel, 2009b).

- The potential to **respond**. Knowing what to do or being able to respond to regular and irregular changes, disturbances and opportunities by activating prepared actions, by adjusting the current mode of functioning, or by inventing or creating new ways of doing things.
- The potential to **monitor**. Knowing what to look for or being able to monitor that which affects or could affect an organisation's performance in the near term – positively or negatively. (In practice, this means within the time

frame of ongoing operations, such as the duration of a flight or the current segment of a procedure.) The monitoring must cover an organisation's own performance as well as what happens in the operating environment.

- The potential to **learn**. Knowing what has happened or being able to learn from experience, in particular to learn the right lessons from the right experiences. This includes both single-loop learning from specific experiences and the double-loop learning that is used to modify the goals or objectives. It also includes changing the values or criteria used to tailor work to a situation.
- The potential to **anticipate**. Knowing what to expect or being able to anticipate developments further into the future, such as potential disruptions, novel demands or constraints, new opportunities or changing operating conditions.

It is straightforward to explain why the four potentials are necessary. If we look at each of them in turn, an organisation is doomed if it is unable to respond when something happens, possibly in the short run and definitely in the long run. This goes even for organisations that are 'too big to fail'. The same argument applies for the potential to monitor. For an organisation that does not monitor what goes on, every situation will be unexpected and a surprise. But being constantly surprised is neither a desirable nor a sustainable condition. The potential to learn is necessary since an organisation otherwise would be restricted to the responses it had at the beginning, with no possibility of changing or improving them. But unless the operating environment is completely stable – and no environment is in the long run – the responses must change and develop over time, which means that learning is necessary. Learning must serve to strengthen or reinforce that which worked well and change or adjust that which did not work well. (Without learning, monitoring will also be limited in the same way as responding.) The potential to anticipate, finally, is needed because an organisation must pay attention to that which is potentially possible even though it may not have happened yet. The potential to anticipate is clearly needed when an organisation is established, just as when a socio-technical system is designed and built. But it is also necessary during the actual operation since the operating environment inevitably is going to change in the future. In the case of nontrivial sociotechnical systems, this is largely due to the fact that the environment includes other organisations that continue to develop and change. In short, an organisation that does not have the potentials to respond, monitor, learn and anticipate will become the victim of unexpected events and their consequences – not least unexpected events of the negative kind. Without the potential to respond, it must passively suffer the 'slings and arrows of outrageous fortune'. Without the potential to monitor, every situation requiring a response will be a surprise, since there will be no forewarning. Without the potential to learn, monitoring will always look for the same signs and signals, and the responses will always be the same. Finally, without the potential to anticipate, everything an organisation does will be constrained by short-term concerns and priorities. While this may suffice for a while, the potential to look ahead and consider or imagine alternatives is a competitive (and evolutionary) advantage.

Two more questions remain to be answered. The first is whether the four potentials mentioned here (to respond, to monitor, to learn and to anticipate) are sufficient or whether other potentials are necessary. This question will be addressed specifically at the end of this chapter after the four potentials have been described and characterised in detail. The second question is whether the four potentials are independent of each other. To this the answer is unequivocally no, as hopefully even the initial descriptions have made clear. The dependencies among the four potentials are, however, important in their own right and will therefore be described in a separate chapter (Chapter 6).

The potential to respond

When something happens, few organisations – and by analogy few individuals – can afford not to respond. There may not be a need to respond to everything that happens, but much of what happens will be above a threshold at which point a response is required. This applies as we literally move through the world, when we walk or when we navigate the road traffic; in this case, an important kind of response is to avoid a potential conflict or collision or to overcome unexpected hindrances on the way – or to make use of a short cut. It happens when unexpected situations arise during our work either of a potentially negative or a potentially positive value – as a threat or an opportunity. Generally, it happens when something unexpected – or for that matter expected – occurs and when the consequences (values) of not responding are less attractive than the consequences of responding. The following examples illustrate the importance of being able to respond.

The Ebola crisis in 2015

More than 11,000 people died from Ebola during the epidemic that erupted in 2014 and lasted through 2015. This was a sixfold increase of victims since the discovery of Ebola in 1976. One factor was that the countries most severely affected by Ebola, Guinea, Liberia and Sierra Leone were unable to detect, report and respond rapidly to outbreaks. The situation was made worse by the fact that the World Health Organization (WHO) had a poor history of responding to pandemics. In 2002, alarm was raised about SARS, which did not become the predicted pandemic. In 2009, alarm was raised again about the H1N1 virus, where establishing large batches of vaccine – as it turned out, unnecessarily – resulted in significant expenses. The WHO was again too slow to declare Ebola an international public health emergency, even 5 months after it had been notified of outbreaks (Moon et al., 2015).

The director general of WHO publicly acknowledged that, with the benefit of hindsight, the organisation could have mounted a more robust response. The director general further promised fundamental changes to the organisation, such as creating a single new programme for health emergencies.

Target porn

On October 15, 2015, people shopping at U.S. superstore Target suddenly heard something surprising over the loudspeaker. Instead of the expected announcements, explicit audio from a pornographic film was blasted out for all to hear and kept on for 15 minutes. This was not an isolated incident but something that, according to local media, had happened at least four times since April.

The intrusion used a weakness in the stores' PA system. After the event, it was realised that an outside caller effectively could take control of the intercom by requesting a connection to a certain extension. Interesting in this context, the staff at the stores did not know how to respond and was therefore unable to stop the unwanted 'announcements'.

Characterisation of the potential to respond

The potential to respond is not a matter of simple knee-jerk reactions but is actually a very complicated affair. The following incomplete description presents some of the major considerations relevant to the potential to respond – and therefore also to how that potential can be established and maintained.

The main issues are clearly *when* to respond and *how* to respond. When looking at responding as a function, there must be some of kind of conditions or inputs that trigger or activate the response.

The input can be a change of the situation or a sudden development that interrupts or disrupts ongoing activities. For example, a new command, an unexpected request, a change in direction (goals) or a change in operating conditions, such as sudden rain during a Wimbledon match. The input can also be an internal change in the organisation, for instance, the outcome of monitoring (an alert).

While it is 'natural' to consider the conditions that trigger a response, it is also important to consider when a response should cease or stop. It is important that the response stops neither too early, before the intended outcome has been effected, nor too late, when a continuation does no good. While the triggering signal must be external to the response, the decision to stop or the stop rule may well be internal to the response, i.e. be part of the way the response is carried out – for instance, part of a procedure. In that sense, responding includes a monitoring of the condition or situation where the response takes place to determine whether the intended effect has been achieved.

The main consequence or result of responding is of course the response itself. The response can be something that has been planned and prepared or something that is developed in the situation. There is always an advantage in having considered the possible events and prepared appropriate responses, but there is also a limit to that. For events or situations that occur regularly, it may be cost efficient to prepare a response, but for events or situations that are irregular or infrequent, responses cannot realistically be prepared. These have to be developed when the event happens. Although this may lead to a delay in responding, the delay will usually be economically acceptable.

Before issuing a response, before beginning an action, there may be conditions that must be fulfilled. It may, for instance, be necessary to request and receive proper authorisation. The Piper Alpha accident is a sad illustration of that; a neighbouring platform continued pumping oil because the manager did not have permission to shut down. The reason for this procedure was the huge cost of such a shut down. Requesting and receiving authorisation may possibly affect the timing of the response in an adverse manner. In other cases, the triggering conditions, the alert or instruction (or clearance), may require clarification or confirmation. In many work situations, this is prescribed and specified, such as in air traffic management.

Another necessary condition could be that the right people are in the right place, that other people are out of harm's way, that the time is right (although this might be considered specifically), etc. Generally, the organisation must be in a state of readiness or in a condition where the response can begin. A good example is the response to emergencies such as major earthquakes, landslides and other natural disasters. The conditions may be obvious and the nature of the response known, but the response cannot be started before certain other conditions have been fulfilled, such as the readiness of materials, teams, transportation and communication.

When the response is started, the presence of specific resources may be required. These resources can be described in generic terms as tools, staff and materials or in detail for a specific response and a specific situation. Competent staff is clearly (often) an important resource. It is easier to substitute one tool for another or make do with other materials than it is to make do without the required competence. Material resources can also be critical; in fighting bush fires, for instance, it is not unusual for the capacity of people and material (including trivial things such as water or chemicals) to be exhausted before the fire has been successfully brought under control.

A response must usually also be managed as it is carried out. Responses are rarely single, ballistic actions or activities, something that is set in motion and from there on takes care of itself. Responses are more often composite or aggregated, may contain several different steps or phases and take place over prolonged periods of time. To manage or control these may require procedures and plans that specify what to do. Examples are the emergency operating procedures that can be found in many industrial and transportation activities, contingency plans, evacuation plans, etc. While the response is being carried out, it may also be necessary for the organisation to maintain some degree of normal functioning, even during an emergency or exceptional operation. Although the criteria for acceptable functioning (the performance envelope) may be modified, routine needs still exist and must be addressed.

Finally, the timing of the response is critical. It is important to begin neither too early nor too late, but it may be equally important not to stop too early or too late. In the first case, the intended result may not be achieved, while in the second case valuable resources may be wasted. The timing or synchronisation of the response can also be critical, particularly if the situation is an unusual one or other considerations play a role.

The potential to monitor

Resilient performance is not possible unless an organisation is able flexibly to monitor both what happens in the operating environment (outside the organisation) and what happens inside the organisation (its own performance). Monitoring improves an organisation's potential to cope with possible near-term events – threats and opportunities alike.

Responding when something happens, when there is an unmistakable event or change, is not too hard. At least it is not difficult to know that something has happened and that it therefore is necessary to do something. But responding after something has happened may be too little and too late. If a situation develops in an irreversible manner, a late response must be different from and probable more extensive (costly and of longer duration) than an early response. To manage something effectively, whichever process or activity it is, requires the ability to respond or react even to minor changes, to notice and recognise trends and tendencies that possibly are too small to represent a real change but may have serious consequences nevertheless. In other words, effective monitoring must be proactive; it must be able to recognise upcoming situations and make use of leading indicators.

While all organisations recognise that is necessary to have the potential to respond, the same does not go for monitoring. In some cases, organisations may be justified in considering the potential to monitor as being of little value. This would be the case for systems that exist in environments that either change rarely (as in the case of geological stable environments), that change in highly regular and therefore predictable ways or in which the consequences of a change are so small that they safely can be neglected (essentially, if the systems are decoupled or very loosely coupled). Some examples illustrate the importance of the potential to monitor.

The Prudhoe Bay oil spill

On March 2, 2006, a leak was discovered in a pipeline owned by BP Exploration, Alaska (BPXA) in western Prudhoe Bay, Alaska. It took 5 days for it to be stopped, and up to 212,000 US gallons (about 5,000 bbl) were spilled over an area of 1.9 acres. The spill came from a 0.64 cm hole in a 86 cm diameter pipeline. The inspection following the spill revealed that the thickness of the pipeline wall had been reduced by more than 70% due to corrosion.

Corrosion of pipelines is a well-understood process, and the level of corrosion can be monitored. A comparable pipeline, operated by Alyeska Pipeline Company, was monitored by the following means: a scraper pig was used every 2 weeks; a 'smart pig' that checked for corrosion was used every 3 years; inspectors made visual inspection by flying over the line each week; a visual inspection by car was carried out every 3 months and finally a manual inspection was carried out every year by walking alongside the entire 1,287 km pipeline.

In sharp contrast, BP Exploration did very little monitoring. The company had determined that the risk of corrosion was low and therefore that there was no need to pig. Indeed, two lines had not been pigged for 8 and 14 years, respectively. Monitoring was instead done by corrosion spot checks on metal coupons stuck into the pipeline supplemented by external ultrasonic spot tests. The company was thus unaware of the actual conditions of the pipeline but simply assumed that it was OK.

In the congressional hearing that followed the oil spill, the president of BP Exploration Alaska stated that the company believed its corrosion control program had been adequate. 'We strive to know the condition of our lines at all times. ... Clearly, in retrospect, this was something we missed'.

Consumer Price Index to measure inflation

A common goal for all governments around the world is to ensure that their countries have healthy economies and in particular that inflation is under control. The preoccupation with inflation is justified by the examples of inflation running out of on control, in extreme cases leading to hyperinflation, that the world has seen in the 20th century. More practically, governments want to keep inflation low in order to reduce the need to spend on various cost of living adjustments. In order to do so, it is important to be able to monitor inflation.

One often-used indicator for inflation is the Consumer Price Index (CPI). The CPI is a statistical estimate based on a sample of representative items whose prices are collected periodically. The CPI is convenient because it is a single number, and it is measured and reported regularly, for instance, every month, but it has several serious shortcomings. One problem is that the CPI mixes different items ranging from food and other consumables to spending on major items such as cars and houses. Some of these items are clearly purchased daily and quickly consumed, while others are only bought rarely, or very rarely, and used for long periods of time. Another problem is that the CPI assumes a constant set of buying habits and does not take into account how these can change over time and be subject to fluctuations that have more to do with geopolitical events, for instance, than the national economy. The various items that make up the CPI have different weights or levels of importance. Determining these weights, and furthermore deciding how often they should be changed to correspond to the actual state of affairs, is difficult and sometimes controversial.

Finally, even if the CPI were generally accepted as a 'true' measure of inflation, there is no agreement on what the response should be, on how inflation can best be controlled. Some countries rely on a monetary policy, some on wage and price control and some on devaluating their currency.

Characterisation of the potential to monitor

The purpose of monitoring is to keep an eye on what happens in the operating environment and in the organisation itself. Most organisations monitor what happens around them because they need to sustain their own existence, to stay alive, so to speak, in their operating environment. As already noted, the best and simplest argument for monitoring is that without it everything that happens will be a surprise. That is clearly not sustainable for an organisation in the long run and probably not even in the short. In practice, an organisation should be as prepared as possible for what can happen and that requires monitoring.

While monitoring of the operating environment, of what happens outside an organisation, is necessary, it is not sufficient. It is also necessary to monitor the state of the organisation itself, to keep an eye on what happens internally. A lack of knowledge of what goes on inside an organisation, not knowing the states or readiness of an organisation, will impede the potential to respond. Knowledge of what happens internally is, however, often neglected or not seen as relevant. A celebrated (negative) example of that is the ignorance in the Airbus company of the wiring problems with the A380. On June 13, 2006, Airbus admitted that 'bottlenecks' in the production of the A380 would force it to postpone customer deliveries by up to 7 months. The co-chair of the company said he had no knowledge of the production problems with the A380 until Airbus made the announcement that Tuesday. At a meeting on June 19, according to a press release, 'Management discussed the problem we are facing with the A380 and the question of how structural changes in the company can help prevent such a problem from happening in the future'. In other words, there was no monitoring, no contingency plan or readiness to respond – and no serious effort to learn.

Monitoring, regardless of whether it is of the exterior or the interior can be based on either indicators or trends. An indicator (from the Latin word *indicō* meaning 'point out') is a signal, a sign or a symbol that represents the current value, magnitude or direction of something. A trend is a general tendency of a course of events measured over time, for instance that a value is growing or diminishing. Where an indicator tells you whether a threshold has been reached, a trend may tell you that a threshold can or will be reached in the (near) future if the underlying developments continue in the current manner.

The result or outcome of monitoring is not just the specific value (of the indicator) or the specific trend, but the interpretation of that. This can take the form of an alarm or an alert, the former being a direct call to action while the latter may start the preparations of a response but not the response itself. A current example, unfortunately, is that a society or a nation can be put on high alert following developments that are seen as a potential threat or something that may be critical for the continued functioning of an organisation. The purpose of monitoring is clearly either to trigger a response or to cause an organisation to change from one state to another, from 'hot standby' to 'operation'. Unfortunate but all-too-ready examples (in 2016) are terrorist attacks in several European capitals or sudden pandemics such as the appearance of the Zika virus.

Unlike the three other main potentials (to respond, to learn and to anticipate), monitoring must take place all the time – although possibly with varying frequency. There may, of course, be special conditions of high alert where monitoring is intensified. A simple example is the monitoring of volcanoes that are in danger of erupting, the monitoring of companies where finances are seen as being in a precarious state, the increased monitoring of a patient in intensive care, or the monitoring of the clock when a deadline (of departure) is approaching. As these examples show, monitoring may change either in frequency or in the indicators that are being monitored – or of course both.

Monitoring may often require specific sensors, equipment or technology, particularly when physical or physiological processes are involved. Monitoring may be done locally or remotely; in the latter case the access to communication technology and transmission channels is essential. Monitoring in many cases relies on people as sensors or interpreters, not least if the focus is on social or organisational processes. One example is the use of opinion polls in the period before an election, customer surveys for a commercial organisation, user feedback (found nearly everywhere, now even in many public toilets), etc.

Monitoring must of course be focused. It must be known what the object or target of monitoring is and especially why. The latter is crucial since a measurement or an indicator that cannot be understood or interpreted is of little or no value – although it will still consume limited resources that could have been used better elsewhere. The way in which monitoring is carried out, the frequency of monitoring, the focus of monitoring (parameters and values), the criteria and thresholds, etc., are all important for effective or at least efficient monitoring to be possible. The control of monitoring is usually based on lessons learned with regard to key performance indicators and thresholds of safe operations.

Monitoring must not only be continuous but must also be given high priority. It is risky to suspend monitoring when a critical situation arises that demands attention. In fact, one could argue that monitoring becomes even more important under such conditions. There must also be time enough to monitor and to do it well. Monitoring is often reduced in frequency when it is seen as providing little of value. If measure after measure is the same, an organisation may come to the conclusion that the measurements are superfluous and therefore change the frequency of monitoring. An unfortunate example is Alaska airlines flight 261, which on January 31, 2000, crashed into the Pacific Ocean north of Anacapa Island, California, killing everyone on board. The probable cause was a loss of airplane pitch control resulting from the in-flight failure of the horizontal stabiliser trim system jack-screw assembly's acme nut threads. The thread failure was caused by excessive wear resulting from Alaska Airlines' insufficient lubrication of the jack-screw assembly. Contributing to the accident were (1) Alaska Airlines' extended lubrication interval and the Federal Aviation Administration's (FAA) approval of that extension, which increased the likelihood that a missed or inadequate lubrication would result in excessive wear of the acme nut threads, together with (2) Alaska Airlines' extended end play check interval and (3) the FAA's approval of that extension, which allowed the excessive wear of the acme nut threads to progress to failure without the opportunity for detection.

A potential risk related to monitoring is to 'jump the gun', to respond preemptively, before a response is actually needed. The advantages of responding in the expectation of a certain development is of course that the situation may not have developed very far and that smaller or fewer corrections therefore will suffice. The possible risk is that a response is made when it is not needed at all or that the response is wrong, both of which can have side effects. Waiting until everything is certain is, of course, playing it safe but may incur a cost. It may also slip to the other side, namely, a response that comes too late and therefore may require more efforts than otherwise.

Indicators

A discussion of monitoring is inseparable from a discussion of indicators. The main purpose of performance indicators is to provide a basis for knowing how an organisation, or a process, performs. For that purpose, it is reasonable to distinguish among three types of performance indicators: (1) lagging indicators, which refer to what has occurred or to organisational states of the past; (2) current indicators, which refer to what is occurring now or to organisational states of the present and (3) leading indicators, which refer to what may occur or to possible organisational states of the future.

- Lagging indicators refer to data that were registered or collected in the past either for the purpose of monitoring or for some other purpose. In the latter case, they were not recognised or used as indicators at the time. They can be used at a later stage, after the fact so to speak, to understand what has taken place. Lagging indicators may also include aggregated data that illustrate a historical development or document a trend. Examples of lagging indicators are event statistics, trends, etc. Lagging indicators are often used as the rationale to adjust functioning following disturbances.
- Current indicators show the state of an organisation at the moment – 'now'; examples of current indicators are production rates, resource or inventory levels, number of aircraft in a sector, number of patients in the waiting

room, fuel reserves, cash flows, etc. Monitoring uses current indicators to adjust performance during operations. This is also known as feedback (control).

- Leading indicators of course are not actual measurements of future states since that is physically impossible. Rather, leading indicators are the interpretations of measurements of current and past states with regard to what may happen in the future. In this way, measurements are used as predictors, rather than as status or performance indicators. Examples of leading indicators are the combined interpretation of available resources, technical status of safety critical components and time available.

Since leading indicators represent an interpretation of possible future developments, there is an irreducible uncertainty. If the uncertainty is deemed too large, warnings or sentinel events may not lead to action. A good example is the difference between tsunami warnings and earthquake warnings. In the first case, the validity of indicators is well established, and people – and authorities – will therefore respond as required. In the latter case, the indicators are more open to interpretation, and people – and authorities – may therefore choose to ignore them. (Additionally, there is a significant cost if actions are taken when they are not necessary – and vice versa.)

The potential to learn

Many organisations treat learning as the proverbial stepchild. But the argument for the usefulness of learning is simple: without learning an organisation will be limited to a given set of responses and will similarly always monitor the same values and conditions. In both cases, the organisation becomes ossified, stuck in established ways of working and therefore unable to adjust to a changing environment. Indeed, even definitions of resilience couched in terms of reactions to threats and disturbances point out that it is necessary to be able to modify the responses because changes in the operating environment otherwise will make an organisation less able to cope with new types of threats.

Learning can more formally be defined as the ways in which an organisation modifies or acquires new knowledge, competencies and skills. Learning does not happen all at once but incrementally builds upon and is shaped by previous knowledge. It is thus better understood as an active process of development than as a passive collection of facts and knowledge. Learning is necessary for individual human beings, social groups and organisations, and the potential to learn is crucial for resilient performance.

The potential to respond and the potential to monitor both depend on the potential to learn, except in rare cases where the operating environments are perfectly stable and perfectly predictable. Efficient and systematic learning from experience requires careful planning and ample resources. The effectiveness of learning depends both on the basis for learning, i.e. which events or experiences are taken into account, and on how the events are analysed and understood.

In learning from experience, it is important to separate what is *easy* to learn from what is *meaningful* to learn. The level of safety, or more precisely Safety-I, is often couched in terms of the number or frequency of occurrence of adverse events. But compiling extensive accident statistics does not mean that anyone actually learns anything. Counting how often something happens is likewise not learning. Knowing how many accidents have occurred, for instance, says nothing about why they occurred or about the situations in which accidents did not occur. And without knowing why something happens, as well as knowing why it does not happen, it is impossible to propose effective ways to improve safety – regardless of whether it is Safety-I or Safety-II.

Safety management practices have traditionally prioritised learning from adverse events (accidents and incidents) both because they attract attention and because they are a cause of concern. In accordance with this logic, it has been assumed that the more serious an event is, the more important is it to learn from it – and also that there is more to be learned. It is easy to demonstrate that both assumptions are wrong, and that they are based on sentiments and values rather than on evidence. The assumptions also seem to overlook the fact that there is a clear inverse relation between the seriousness and frequency of events, so that serious events (major accidents and disasters) are very infrequent. There are therefore few occasions to learn (cf., the discussion of 'snapshots' in Chapter 1). Learning can clearly benefit from broadening the basis to comprise not only smaller accidents and near misses but also events that are not classified as accidents at all. Since the number of things that go well, including near misses, is several orders of magnitude larger than the number of things that go wrong, it makes good sense to try to learn from events that are representative in terms of their frequency rather than just from events that cause concern due to the severity of outcomes.

If learning focuses on events that happen frequently, on everyday work and activities, it must take place on a continuous basis rather than being a reaction to a single, serious event. In an organisation with a good 'learning culture', everyone adopts a learning mode and accepts it as a natural part of everyday work. A good 'learning culture' is based more on the discovery of good – and bad – practices and on gradually absorbing their consequences than on a targeted analysis of specific events. It is learning on a breadth-first rather than a depth-first basis.

Saddleback fatality learning review

Investigations into and reviews of accidents and incidents usually try to develop an understanding of what went wrong, in accordance with a Safety-I perspective. Indeed, reviews of events are mostly, if not exclusively, motivated by adverse outcomes, hence limited to events where something presumably has gone wrong, even though there may be as much – or possibly even more – to learn from something that has gone well. The Saddleback Fatality Learning Review, however, is different because it emphasises the importance of learning rather than of merely identifying causes.

In this event, which happened on June 10, 2013, three firefighters were constructing a fire line around a tree that had been struck by lightning, within the South Warner Wilderness of the Modoc National Forest in California. At approximately 17:00 h, a limb fell out of the tree and struck one of the firefighters. The two other firefighters started CPR and called for emergency evacuation. The helicopter, which was stationed a 55-minute flight away, landed at the place of the accident at approximately 18:19 h. The injured firefighter was transported to the nearest hospital but did not survive, despite all efforts to resuscitate him.

The learning review is interesting because it intentionally refrained from drawing conclusions in the traditional way of accident reports. The review instead tried to 'empower readers to explore, question, and learn' by giving them information that would allow them to 'make their own determination of why the decisions and actions at Saddleback made sense to those involved'. It did, nevertheless, suggest some conclusions, although they differed from the usual litany of causes and performance-shaping conditions. Based on eight focus groups with firefighters, the learning review team concluded that the event had elements (conditions, decisions and actions) that were recognised as common, and that it therefore could be seen as normal work. The event was unique not in terms of the number and variety of conditions but in terms of how they happened to combine in an unexpected manner.

A failure to detect patterns

On February 6, 2014, General Motors (GM) recalled about 800,000 of its small cars. The reason was that faulty ignition switches could shut off the engine during driving and thereby prevent the air bags from inflating. Cars could also sometimes stall at high speeds on highways, in dense city traffic and while crossing railroad tracks. The company continued to recall more cars over the following months, and in the end nearly 30 million cars had to be recalled worldwide.

While GM apparently was slow to learn, or slow to take the consequences of the reports, this case also points to a more interesting and in some sense more significant failure to learn. In the USA, traffic safety is overseen by a federal safety regulator, the National Highway Traffic Safety Administration (NHTSA). The officially described mission of the NHTSA is to 'save lives, prevent injuries, and reduce vehicle-related crashes'. It was found that the NHTSA since February 2003 had received an average of two complaints a month about potentially dangerous shut downs in GM cars, altogether more than 260 complaints, but the regulator had repeatedly responded to the complaints by saying that there was not enough evidence to warrant a safety investigation. For some reason the NHTSA failed to see a pattern in the many complaints and there-fore failed to learn. Perhaps it is more correct to say that it failed to recognise an already known pattern in the data – thus a classic case of What-You-Look-For-Is-What-You-Find (WYLFIWYF). Learning clearly requires more than a recognition of what is already known.

In the case of GM itself, it was determined that the fault had been known to the company for at least a decade before the recall was issued. The recall only happened because of a determined attorney who sued GM on behalf of the family of a woman who had died in a crash. In addition to recalling more than 30 million cars, GM paid compensation for 124 deaths.

Characterisation of the potential to learn

The essential basis for learning is the responses that have been made, both those that succeeded and those that failed. Another important source of learn-ing is the organisational experiences, i.e., how the organisation has developed over time, how well it has performed, etc. A primary focus for learning is of course the relations between interventions (responses) and outcomes from the responses, or more precisely which means have been effective towards which ends. One critical issue here is how soon the consequences of the responses can be expected to show. Some may be immediate or quickly follow the response, while others may take a long time, such as changes to procedures, attitudes and the like. Another issue is how closely or 'directly' outcomes follow from interventions. In some cases there may be an indispensable cause-effect relation (or at least an inferred one); in other cases the links or connections may be less certain or perhaps even circumstantial. The output of learning, the 'lessons learned', may also be expressed in many different ways and have many differ-ent manifestations. They can range from redesign of equipment and tools, to revision of procedures, changes to rostering and work allocation, retraining, re-vision of goals and priorities, revision of indicators and measurements or even courageous attempts to change the organisational culture.

In order to learn, several things are needed. First of all competent staff, in-cluding competent leadership, since learning is still basically a human activity (and is likely to stay so for the foreseeable future). People are needed to collect data and information, to analyse them, to draw the conclusions and to formu-late how the lessons learned can best be brought into practice. Learning may also require some equipment, particular some kinds of IT. And it definitely requires time and money, in other words, learning must be prioritised by the organisation.

Another important issue is how learning takes place and how it is controlled. This is essentially an issue of how important an organisation considers learning to be. Does learning take place continuously as an integral part of the daily routine, or is it something that is done whenever there is an obvious or un-avoidable need (usually meaning that something has gone seriously wrong)?

How learning takes place is also a question of whether there is an overall strategy for learning, whether there is the necessary organisational support, etc.

Learning, finally, depends very much on time, first in the sense that learning itself takes time. Analysing experience and drawing the appropriate conclusions is only the first part or step of learning. Nothing will be learned just by summarising experiences. The experiences must somehow be turned into practice, ways must be found to make sure that the appropriate changes are made to the organisation and the way it works.

Another temporal issue is that learning can only take place when the responses have had time to work and when the results become manifest. There may be huge variations with regard to time and duration. Some changes (responses) can be assumed to have nearly immediate effects (such as a change to equipment), while others may take longer to show themselves – either because the change takes time or because it must await a similar occurrence.

In the worst case, an organisation may not learn anything at all. This happens if the primary concern is on short-term productivity and efficiency, which encourages a 'fix-and-forget' tactics. When people are faced with a problem, do they strive to fix it immediately and forget about it, do they fix it right away and report it or do they fix it, report it and try to learn from it? Fixing and forgetting is a short-sighted solution that may take care of pressing problems in the short run, for instance when there are significant external pressures towards productivity and accessibility. If the time perspective is a little longer, and if the approach is tactical rather than opportunistic, 'fixing and reporting' is another solution. Fixing and reporting tacitly assumes that someone will take care of the reporting and learn from that. This hope is not always realistic, and even if learning takes place, it is removed from the actual situation where something happened. The best solution is therefore fixing, reporting and learning. Only the latter approach aligns with a preventive view of safety.

Prerequisites for learning

Looked at from a more theoretical or academic point of view, three conditions must be met for learning to take place. The first condition is that there must be an opportunity to learn. This means that there must be a recognition that the actual situation – or the actual outcome of an activity – is so different from the expected situation or outcome that it is necessary to understand why in order to do something about it. This is naturally the case when the outcome is serious (has a large magnitude) and when the value is negative (an adverse outcome). However, it should also be the case when something happens frequently and the outcome is acceptable since this can say something about the nature of everyday work. Daily practice quickly establishes a threshold at which learning takes place, although that threshold is usually too high. In other words, an organisation – or person – must recognise an opportunity to learn or be forced to learn when current performance leads to unacceptable results.

A second condition is that there must be a reasonable degree of similarity between the situations where the need to learn is acknowledged. The explanation for this is straightforward. If the situations are different or dissimilar, then learning has to be specific to the situation, meaning that a transfer of learning (of knowledge, of competence) to other situations is impossible. Each situation is unique, and each lesson is unique. In practice, there is nearly always some commonality in the operating environment, hence in the aetiology of events. It is clearly easier – and also better – to find similarities among situations and to learn something that is generic or has general value. Even though the similarities may be perceived or interpreted rather than real, the presence of similarities is nevertheless a condition for learning. No organisation can continue to exist in an environment where each situation is unique.

The third condition for learning is that there is an opportunity to verify or confirm that something has been learned. Learning is essentially a change in behaviour or performance and not merely a change in knowledge. We may feel that we have learned something, and others may claim that they have learned something, but unless that can be demonstrated by differences in performance, it remains a postulate. In order for such changes to be noticeable or observable, it is necessary for the same, or a sufficiently similar, condition to occur again. The third condition interestingly enough means that it is very difficult to confirm that anything has been learned from a serious accident or a disaster. There may be consequences from the effort to learn, such as changes to the organisation and the ways it works (procedures, equipment, tasks, responsibilities). Issuing a new procedure is, in some sense, evidence of learning, but until and unless a situation arises where the new procedure has to be used, we cannot with any reasonable conviction claim that anything has been learned.

The three conditions or prerequisites for learning can be illustrated as in Figure 4.1. The X-axis represents the degree of similarity between situations or events, while the Y-axis represents the frequency of occurrence. The similarity is highest among everyday events, and these occur most frequently. At the other end of the scale, the similarity is lowest for serious events such as accidents, and these are likewise events that happen infrequently. (This is partly because we take steps to insure that accidents do not happen while we try to facilitate that everyday events happen.) As Figure 4.1 shows, the 'consequence' is that there is more to learn from everyday events than from accidents. This is, of course, consistent with a Safety-II perspective.

Failing to learn

Although it should be obvious that learning is necessary, several studies have shown that organisations sometimes not only fail to learn but do so deliberately. To be more precise, people at leading positions in organisations and in charge of learning (or with the power to control learning) sometimes fail to do

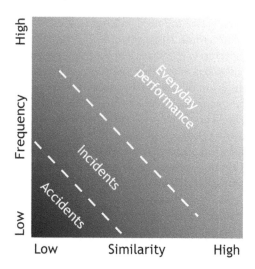

Figure 4.1 Prerequisites for learning.

so (Baumard and Starbuck, 2005). There can be many different reasons for this, but some of the more common are the following.

- Learning is inconvenient at the time – or in general – because it is seen as conflicting with other priorities, typically productivity.
- Learning is nice, but it is time consuming and expensive. This is the cost and effort argument.
- Learning is risky, in the sense that it exposes the organisation and gives the impression that it is deficient. Learning is an admission of being weak and inadequate.
- Learning is a threat, not least against a person's (or leader's) own position. It is a weakness that may be used by others in a power struggle, a cinch in the armour.
- Learning is dissonant; it goes against the values of the person or the organisation.

The potential to anticipate

That monitoring is essential for an organisation hardly needs any argumentation. An organisation must pay attention to what goes on both in the operating environment and in the organisation itself and must be able to make sense of it. Without that, it will be unable to function efficiently for long. But it seems to be less obvious that it also is useful for an organisation to look to the more distant future to anticipate possible events, conditions, threats and opportunities that may strengthen or weaken its continued functioning.

Monitoring looks at that which is within the event horizon, within the scope of current operations and activities. Monitoring looks as that which exists and changes, provided that it can be perceived or measured or otherwise noticed. Monitoring is about watching, observing or checking something to see whether it is changing in a way that requires a response or an intervention. Anticipation 'looks' at that which is beyond the event horizon, either something that lies further into the future or something that has no immediate relation or impact on the organisation's primary activity. Where monitoring is *looking* at something, anticipation is more like *thinking* about or *imagining* something.

Such anticipatory thinking is fundamental for individuals, for groups, for organisations and for societies large and small. It is necessary to be prepared for what may happen in the future either as threats or as opportunities. Just as an organisation must try to make sense of the present – and of the past – it must also try to make sense of the future. The problem is, to quote Norbert Wiener (1954), 'The present is unlike the past, and the future is unlike the present'. The future is uncertain, but it is second nature to humans – and therefore also to organisations – to try to reduce that uncertainty as far as possible. Humans can face uncertainty in many ways, by being fatalistic, by pessimistic resignation, by eternal optimism, by determinism, etc. All of these may give peace of mind and allow a person to pull through, but none of them has any direct effect on the potential to anticipate.

There are two other ways an organisation looks into the future and have a superficial similarity to anticipation. The first is the planning that guides an organisation's behaviour. The essence of planning is to think about and arrange the details of activities before they take place. Planning therefore by necessity refers to the concrete or actual whereas anticipation refers to the hypothetical or potential. Another important difference between planning and anticipation is that planning is synchronous where anticipation is asynchronous. Planning is synchronous because it is part of ongoing activities and indeed necessary for an organisation to be able to act at all. Although planning can have a short horizon or a long horizon, as in tactics and strategies, the primary purpose is the preparation of future activities based on current activities and the current situation. The purpose of anticipation is not to support the current activity but to imagine alternative scenarios and to think about what potentially can be done in a completely different situation.

The second activity that looks into the future is risk assessment. The purpose of risk assessment is to identify, ahead of time, conditions or events that may pose a threat to an organisation in the sense that it may prevent it from carrying out is functions or lead to unacceptable losses of life, materials or resources. A risk assessment is useful for organisations and systems that are tractable. This means that the principles of functioning are known, that the descriptions do not contain too many details and can be produced relatively quickly and that the organisation and its operating environment are so stable that descriptions remain valid for a long time. Many present-day organisations unfortunately do not meet these conditions but are instead intractable. For such organisations the principles of functioning are only partly known, descriptions contain (too) many details and take a long time to produce, and the organisations or their operating environments keep changing so quickly that descriptions must be frequently updated. (In extreme cases, organisations may change faster than descriptions can be produced. Descriptions will therefore always be incomplete, and the organisations as a result underspecified.) In such cases, traditional risk assessment methods are inadequate, if not downright inappropriate. A risk assessment differs from anticipation by being constrained by the available description of the organisation's functioning. The description is systematically analysed and evaluated, but unlike anticipation, risk assessment cannot go beyond that and consider disjoint alternatives. Another difference is that risk assessment is focused on what may fail or go wrong. There are no corresponding methods that can be used to look for future opportunities even though this rightly ought to be considered as important as the search for threats.

Anticipation and models of the future

Since anticipation is the expectation of what will or may happen in the future, it depends very much on how we think about the future, which inevitably also means how we think about the present and the past. More specifically, it depends on the assumptions we make about how things happen. In practice, we can distinguish several typical views or models.

- The simplest form of anticipation relies on recognition, relating to similarity or frequency. This corresponds to a 'mechanistic' view, where the future is a 'mirror image' of a repetition of the past. In other words, if a situation matches something that has been experienced in the past, either based on similarity or frequency, it is assumed that it will develop in the same way in the future.
- A more elaborate form is by extrapolation where the unknown future is inferred from the known past, specifically easily identifiable trends and tendencies (real or spurious correlations). This corresponds to a 'probabilistic' view where the future is described as a (re)combination of past events and conditions.
- The final form of anticipation is by a deliberate construction of possible future situations based on an understanding of the past and the present and relying on deduced or inferred principles of how events develop. This corresponds to a 'realistic' view that acknowledges that the future has not been seen before. Anticipation may be based on a combination of what is already known, but it often involves variability and adjustments that have been seen as trivial hence insignificant in the past.

While global warming is the example par excellence, less cataclysmic examples can easily be given.

Cheap travel (SNCF)

What happens if a train company, such as the French National Railways (SNCF) starts a campaign for cheap travel and offers a very cheap ticket (one Euro) in combination with people returning from a bank holiday? When the question is asked this way, it is hardly surprising that the answer is an increase in the number of passengers, which of course was the intention. But the more interesting question is what consequences a larger number of passengers will have for daily operations.

The case that gave rise to this question was the derailment of a regional express train on an incorrectly positioned turnout upon its arrival at a station. One factor contributing to the derailment was that the signaller, who was responsible for setting the turnout in the right position, also was in charge of ticket sales. Since there were customers waiting at the ticket office, he tried to save time and therefore did not position the turnout 'by the book'. In addition to the exceptional workload, the signaller had agreed to work a double shift because his replacement had not arrived. There were, as usual, other circumstances complicating the situation. The interesting question here is the failure of the company to anticipate that reduced ticket prices would lead to more passengers, which in turn would lead to more work for the staff, specifically the single person who was in charge of both ticket sales and signalling. The focus on the intended goal, an increase in the number of passengers, led to a disregard of other possible consequences.

(I am grateful to Christian Neveu, Chef du Pôle Facteurs Organisationnels et Humains, SNCF, for letting me use this example.)

Cloned meat (US)

On January 15, 2008, the U.S. Food and Drug Administration (FDA) declared that food from cloned animals was safe. This removed the last obstacle for producers who for more than 4 years had sought approval to sell such products at grocery stores. There had originally been a tentative declaration from the FDA that food from cloned animals was safe, but it had – not unexpectedly – been met by criticism from both consumer groups and concerned scientists. In January 2008, the FDA announced that further studies had confirmed the earlier decisions.

Without going into a discussion of whether or not the FDA's decision was correct, it is interesting to note that the FDA's chief food safety expert issued the following remarkable statement to reporters: 'It is beyond our imagination to even find a theory that would cause the food (derived from clones) to be unsafe'. Perhaps the chief food safety expert was right, but perhaps it rather demonstrated an inability to anticipate and a lack of imagination.

Characterisation of the potential to anticipate

The foundation for anticipation is the realisation that it is necessary to think beyond the current situation or moment. It is the realisation that the future is uncertain and that it therefore is necessary to consider which preparations might be necessary. It is the acceptance of the need to be prepared.

Anticipation must, however, be based on something. It is related to a vision of what an organisation would like to be and where it would like to be at some point in the future, but a vision is not enough. A corporate vision may serve to guide or control the anticipation, but a corporate vision is not by itself anticipation. It is rather a programmatic statement that, in the best of all possible worlds, becomes a self-fulfilling prophecy.

The output from anticipation, what anticipation produces, so to speak, is a set of potential foci, critical concerns or priority areas. The output represents an organisation's ideas about how future developments and conditions could affect its existence and performance. To provide this may require research and exploration (but focused) or the development of specific capabilities and resources. That is why anticipation may be difficult to do continuously (except as a constant sense of unease) and instead takes place whenever it is felt necessary. The precondition for anticipation is thus a more or less distinct sense of unease, a feeling that the situation is no longer stable, predictable or certain. The 'sense of unease' may be triggered by events that are not directly related to what the organisation does but are seen to have some bearing even though there is no formally recognised 'causal pathway'. A complacent organisation does not consider anticipation necessary because it thinks it is insulated from all but a limited set of external events. An example is the FIFA in the years 2010–2015, during which period the organisation considered itself impervious to what happened around it. Or the Republican party in the USA before the 2016 primaries began. A 'resilient' organisation anticipates because it realises that there is no such insulation, even though it may not be possible explicitly to explain how things hang together.

Anticipation is an art rather than a science and is akin to requisite imagination (Westrum, 1993). It may therefore be difficult to define specific resources that are required for a successful realisation, other than what one could call a 'think tank', a group of people (internal, but perhaps also external) that have the freedom – and the resources – to speculate about what could possibly happen. Such people should by virtue of their assignment be unhampered by the usual organisational (and cognitive) constraints – at least when they are engaged in anticipation. The most important resource is arguably time. Anticipation runs at unknown speeds, with unpredictable starts and stops. There can be a request to think about the future, but it would be a bad sign (in terms of corporate vision) if there also were a strict deadline for when the results should be ready.

While anticipation should be unconstrained, it should not be uncontrolled. It must still be reasonably productive, hence in some sense be controlled or guided, or at least scrutinised from time to time. One way of checking and guiding anticipation is by referring to the commonly recognised threats and opportunities, as well as to the threats and opportunities that come from the corporate strategy. Another control is the corporate vision, the general idea or image of what the company or corporation would like to be some years down the line. (The most unfortunate example is BP's vision set out in the Baker report: global leader in safety. This will be described in Chapter 5.)

Issues in the potential to anticipate

The dilemma of anticipation is that it cannot be known with certainty, which aspects of a problem can be ignored and which should be attended to. This has in the literature been described as an exploration-exploitation dimension (March, 1991). Exploration includes things captured by terms such as search, variation, risk taking, experimentation, play, flexibility, discovery and inno-vation. Exploitation includes such things as refinement, choice, production, efficiency, selection, implementation and execution. Organisations that engage in exploration rather than exploitation are likely to find that they suffer the costs of experimentation without gaining many of its benefits. They exhibit too many undeveloped new ideas and too little distinctive competence. Conversely, organisations that engage in exploitation to the exclusion of exploration are likely to find themselves trapped in suboptimal equilibria. As a result, main-taining an appropriate balance between exploration and exploitation is a pri-mary factor in system survival and prosperity.

Anticipation may be hampered by rigidities in institutional beliefs, distracting decoy phenomena, neglect of outside complaints, multiple information-handling difficulties, exacerbation of the hazards by strangers, failure to comply with regulations and a tendency to minimise emergent danger. The antidote is to keep an open mind, although that is easier said than done. It is dangerous for an organisation to think that it knows better than outsiders the hazards of the situations with which they are dealing.

Other potentials?

When the four potentials are put forward, two questions immediately come to mind. The first is why there are four potentials rather than three or five or some other number. The second is why the four are respond, monitor, learn and anticipate rather than four other potentials. Both questions can rather easily be answered.

The reason there are four potentials is pragmatic rather than logical or deduc-tive. The four potentials proposed here can easily be recognised in many descrip-tions and analyses of events, and the four taken together seem to be sufficient without being redundant. In relation to the definition that an organisation's performance is resilient if it can function as required under expected and un-expected conditions alike, it is clear that none of the four potentials can be dispensed with.

An organisation that is unable to respond is going to fail, possibly in the short run and definitely in the long. Responding must, however, be a simple selection from the same set of responses no matter how large it is, unless the operational environment never changes. The responses must perforce develop over time, which means that the organisation must be able to learn. Indeed, the ability to acquire new skills and knowledge or modify existing ones is practically a definition of learning. But responding cannot be effective unless it is supported by monitoring. Without monitoring an organisation must constantly be primed for any of the possible responses. That is clearly not practically possible; neither is it reasonable from an economic or productivity point of view. Monitoring must also be adjusted based on experiences in the same way that responding is, i.e. it must be based on learning. Together the three potentials of responding, monitoring and learning may enable an organisation to 'muddle through', perhaps even for a considerable time. The organisation's performance may meet the criteria for being safe – in the Safety-I sense – and perhaps also for being effective. But the organisation does not meet the criteria for being resilient because it is unable to prepare itself for what may happen beyond the current situation. It is as necessary to imagine the future as it is to analyse the past. In order to do the former, an organisation must have the potential to anticipate. There is a clear 'evolutionary' advantage in being able to prepare for something that could be possible, although it has not happened yet and although it may never happen. Anticipation may not be needed, although it could still be useful, if the operating environment is stable in the sense that new developments or surprises are unlikely. But if the operating environment changes during the lifetime of an organisation, then anticipation clearly becomes necessary. Although thinking about the future in the short term may appear to be non-productive, hence a cost, that is not the case in the long term. Those who cannot remember the past are condemned to repeat it, as George Santayana pointed out. And those who do not learn from the past are condemned to fail in the future.

Accepting that the four potentials are necessary, one is justified to ask whether also they are sufficient or whether a fifth or a sixth potential may be required. Three obvious candidates are the potential to plan, the potential to communicate and the potential to adapt.

Planning is necessary for an organisation to function since planning famously provides the structure of behaviour (Miller, Galanter and Pribram, 1960). Planning is however not limited to resilient performance but applies to any kind of performance, to everything an organisation does regardless of whether it is short term (tactical) or long term (strategic). Planning is required for an organisation's existence rather than for resilient performance.

Communication is in a fundamental sense necessary for any organisation and any system. Communication is the ability to transmit information, both to receive information about what goes on inside and outside the organisation and to send information in order to exercise control. Communication can therefore be seen as necessary for responding, monitoring and learning and probably also for anticipation. But the potential to communicate does not directly contribute to resilient performance in the same way as, e.g. responding does. Explicit communication is necessary for an organisation to coordinate how the various parts function, but communication, like planning, is required for an organisation's existence rather than for resilient performance.

A third contender is adaptation. The ability to adapt is undeniably important for an organisation, and it has in recent years become *de rigeur* to talk about complex adaptive systems. But adaptation is a composite rather than a basic potential. A system that is adaptive can adjust or modify itself, or rather adjust the way it functions, based on experience. Adaptation can therefore be seen as a combination of the potential to learn and the potential to respond and possibly also of the potential to monitor. Adaptation is therefore not a basic potential.

The above arguments obviously do not rule out that there at some time may be a genuine need for a fifth potential, and there is nothing in the concept of resilient performance or in the principles of the Resilience Assessment Grid (RAG) to prevent that. Regardless of the number and nature of the resilience potentials, it is important that resilience is thought of as the expression of the resilience potentials rather than as a unitary quality or potential of an organisation. The four resilience potentials must be described as functions rather than as components and be seen as constituting a whole by including an account of how they depend on each other, on how they are coupled. This will be done in Chapter 6. Before that, Chapter 5 will explain how the RAG can be used in practice.

RAG – the resilience assessment grid

It is clearly essential that an organisation be able to perform in a manner that is resilient and hence be able to suffer the 'slings and arrows of outrageous fortune'. Yet, it is inappropriate to treat resilience as a monolithic concept: neither resilience nor resilient performance can be managed or controlled directly. If, on the other hand, resilient performance is accepted as an expression of an organisation's abilities or potentials, then resilient performance can be managed indirectly through the potentials for resilient performance. It cannot, of course, be taken for granted that an organisation's potentials will always be realised when the need arises, but an organisation that has them will be more likely to perform in a way that is resilient than one that does not. Conversely, it is pretty certain that an organisation that lacks these potentials will be incapable of resilient performance.

Fundamental requirements to process management

Resilience engineering is basically about how the four potentials can be managed – not just one by one but together (more about this in Chapter 6). In order to manage something, whether it is how an organisation performs, how something is produced, how people or goods are transported from A to B, etc., three things are necessary. It is first necessary to know what the current situation or status is (the current 'position'). Then it is necessary to know what or where the target is, i.e., what the desired future state of an organisation or system is and when it should be reached. It is finally necessary to know how to bring about a change from the current position or situation towards the target – how to 'move' an organisation in the right direction and at the right speed.

As a literal illustration of this metaphor, consider steering a ship that sails from port to port. In order to do so, it is necessary to know the current position (*where* the ship is now), the destination (*where* the port of arrival is and possibly *when* the ship is expected to arrive) and finally *how* to steer the ship so that the difference between the current position and the destination is reduced in an orderly manner. The lack of either of the three types of knowledge makes safe steering impossible. The consequences of not knowing the current position were dramatically illustrated by a naval disaster in 1707. On this occasion, the Royal Navy lost four ships and about 1,550 sailors, when they struck the rocks of the Isles of Scilly. The weather was bad and the fleet mistakenly thought they were just west of the island of Ushant (or Ouessant), some 184 km to the southeast. The consequences of not knowing the target were illustrated by Columbus' first voyage to what he thought was part of the Asian continent, the Indies, in 1492. He did find land but, as we now know, it was not the Asian continent but the American. And finally the problems caused by not knowing how to get from the current position to the target, even when both are precisely known, was illustrated by the Apollo 13 mission in April 1970, at least during the 6 h between the explosion and the time when the revised flight plan was agreed (Table 5.1).

In the case of managing the movement of a physical system the position is literal, namely, the geographical position. In the case of management, the 'movement' of an organisation, for instance, changing attributes such as safety, quality or 'resilience', the 'position' is figurative. A good example is the so-called 'safety culture journey' used by many industries (Foster and Hoult, 2013). The journey begins by determining or 'measuring' an organisation's current safety culture using a five-level description, such as the one shown in Figure 5.1. The purpose of the 'journey' is to change the organisation's level of safety culture, for instance, from being 'reactive' to become 'planned'. The safety journey uses the metaphor of a physical journey, but neither the 'positions' nor the means of 'travelling' from level to level are clearly defined – and perhaps not even clearly understood. While the metaphor is popular in many industries, one basic shortcoming is that the 'position' is very difficult to measure. A further problem is that the organisational target is relative rather than absolute because it depends on what other organisations do or achieve. There is finally very little concrete actionable knowledge about how to 'move' or change an organisation, the confident claims of safety culture aficionados notwithstanding. (This issue will be discussed also in Chapter 7.)

Table 5.1 Consequences of lack of knowledge

	Columbus	Scilly naval disaster	Apollo 13
Knowing current position or point of departure	Yes	No	Yes
Knowing target position or point of arrival	No	Yes	Yes
Knowing how to steer from current position to target	Yes	Yes	No
Consequence	Arrived in the wrong place	Major naval disaster	Uncontrolled drifting

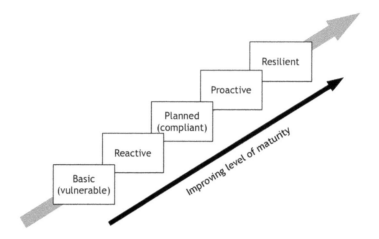

Figure 5.1 The 'safety culture journey'.

British Petroleum and the Baker report

After the accident at British Petroleum's Texas City refinery on March 23, 2005, no less than six investigation reports were produced – with six different sets of recommendations. One of these, the so-called Baker Panel Report, so named because the panel was led by former U.S. Secretary of State James Baker III, examined the corporate safety culture and the safety management systems. The report gave ten recommendations, of which the last was this:

> BP should use the lessons learned from the Texas City tragedy and from the Panel's report to transform the company into a recognized industry leader in process safety management.

This is obviously a recommendation to make a change. The starting position was the company's unrecognised but sorry state of process safety management when the accident took place. The target was to become 'a recognized industry leader', which was not well defined either. The report, unsurprisingly, did not provide concrete advice on how to become a 'recognized industry leader'. Despite the nice sounding words, the recommendation was useless in practice. Five years later, the company's Deepwater Horizon drilling rig exploded in the Gulf of Mexico, resulting in the largest accidental marine oil spill in the world and the largest environmental disaster in U.S. history.

Measurement or assessment?

Measurements are often used to present an organisation's position in relative rather than absolute terms, for instance, by comparing it to a standard, a regulatory norm, an industry average, etc. The purpose of the Resilience Assessment Grid (RAG) is, however, to rate or measure the potentials for resilient performance in order to manage them rather than to make a comparison. It therefore makes little sense to compare a given assessment to some arbitrary standard or even to another organisation. But it does make sense to compare it to an earlier assessment of the same organisation since it in that way serves as its own reference. Repeated assessments make it possible to keep track of how an organisation's 'position' changes and whether it is in the intended direction and with the intended 'speed', to the extent that such a statement is possible.

Chapter 4 has argued that there is only limited value in assessing an organisation's performance using the four potentials wholesale, as single qualities or dimensions. An assessment should instead look at the operational details of the four potentials, at the concrete operations or functions that are necessary for an organisation to be able to respond, to monitor, to learn and to anticipate. Unlike 'resilience', it is easy to see how the potential to respond, for instance, comprises a number of specific operations or functions and how each function can be addressed by one or more diagnostic and formative questions. Questions are diagnostic if they characterise the potentials in operational terms that either can be answered directly or be rated with regard to the extent of agreement or disagreement. Questions are formative if they together with the answers can be used as the basis for concrete activities – interventions or improvements – to develop specific operations or functions in an intended direction – keeping in mind the need to maintain an overview of the dependencies and interactions among the detailed functions and the four potentials. An identical argument can be made for the three other potentials.

'Six honest serving-men'

It is usually not a problem for people with enough practical experience to think of questions that precisely address critical aspects of their work and organisation. Indeed, it is in many cases part of their job, even though it may come to be done routinely (indirectly) rather than diagnostically (directly). Since such questions necessarily must be concrete and specific for an organisation or a type of activity, it is not realistic to develop a set of generic questions that can be applied to any kind of organisation or system. But it is feasible to propose a common set of questions that can serve as the starting point for developing the diagnostic questions for a specific organisation. (How this can be done is described towards the end of this chapter.) This will not least be helpful for people who are relatively inexperienced in conducting these kinds of assessment.

Proposals for generic questions to help in gathering information were put forward in classical rhetoric more than 2000 years ago. Boethius, a Roman senator and philosopher, recommended 'seven circumstances', namely 'who', 'what', 'why', 'how', 'where', 'when' and 'with what'. In the 19th century, the idea of Three Ws (What? Why? What of it?) became popular, and in the beginning of the 20th century Rudyard Kipling wrote about the 'six honest serving-men' – five Ws and one H – What, Why, When, Where, Who and How.

In the case of assessing the potentials for resilient performance, it is clearly important *what* the operations are and *how* they are carried out. This includes issues such as the appropriateness of *what* is being done (or potentially can be done) vis-a-vis the actual conditions whenever it is necessary to do something and *how* it is being done – in terms of intensity or capacity, for instance. The *when* refers to the timing of what is being done – the onset and the duration. The *how* may similarly refer to whether something is done in the right way, given that it is the appropriate thing to do. The appropriateness naturally leads to the question of *why* the specific operation or function was chosen. The *who* is the question of the people – or increasingly the technology – needed to ensure that the potential is fulfilled or implemented. The *where*, finally, is whether the focus or target of what is being done is the right one.

Assessing the four potentials

The following sections present examples of diagnostic questions that can be used to assess the details of each of the four potentials – the subpotentials. The diagnostic questions are derived from characterisations of the four potentials that address concrete and practical issues. The answers can therefore be used to produce a detailed characterisation of each of the four potentials as well as of an organisation's overall potential for resilient performance. Each of the four potentials is illustrated by a set of generic questions. These have for obvious reasons been formulated in general terms without referring to a specific domain and are therefore not intended to be used directly. In order to have diagnostic value, the questions must be revised to capture what is important for a specific organisation. Neither should the number of questions in each set be seen as a norm. Depending on the needs in an actual case, a larger or smaller number may be appropriate. For each potential, a set of actual diagnostic questions illustrates how the generic questions have been applied in four specific domains.

The potential to respond

No organisation, organism or system can exist for long unless it can respond to what happens. The responses must be both relevant and effective – they must be appropriate for what happens, and they must contribute to bring about the right outcomes before it is too late. Since no organisation has infinite resources, responses can only be prepared for a limited number of events or conditions.

One diagnostic question therefore concerns the conditions or events to which the organisation is able to respond.

Every organisation considers a set of 'regular threats', events or situations for which it is cost-effective to be prepared to protect the organisation (Westrum, 2006). Some organisations also consider the possible events or situations that with some regularity may constitute an opportunity, e.g., seasonal variation. For other more 'irregular' events, there may be a general but no specific readiness. An organisation's list of regular and/or irregular events can be based on tradition – euphemistically called the 'collective experience of the industry' – such as commonly used risk scenarios, unacceptable failures, safety cases or past successes. It can be based on legal requirements (directives) or prescribed by regulatory requirements. It can be based on industry standards, system design, expertise, risk assessment, market analyses and so on.

The event list must obviously be revised every now and then to reflect both the 'lessons learned' and changes in the operating environment. These revisions are usually reactive and incremental and are made, for instance, when the organisation has been unable to respond and wants to ensure that this does not happen again. But learning from failures is the bare minimum, and it is clearly better – in every sense of the word – if the revisions are made regularly and systematically, using foresight as well as hindsight.

Another important question is whether the prepared responses are appropriate and/or adequate. The most important sources for selecting responses are unquestionably experience and tradition. The tried and trusted is often preferred because it presumably involves little risk, and if something has worked in the past, it will likely be used again. An organisation that does not have sufficient experience itself may imitate or 'import' what others have done. (The need to emulate the successes of others is occasionally so strong that even irreconcilable differences between two organisations are disregarded.) Responses may also be based on assumptions or hypotheses about how the world works, expert advice, etc. For events that occur regularly, the responses can gradually be refined and calibrated. But for events that occur rarely, such as large accidents or other major upheavals, the effectiveness of the planned responses is unknown until the actual situation occurs.

It is also necessary to ask how effectively a response can be implemented. Will an organisation always be ready to respond? Will people, equipment and materials always be ready when needed? Some events, such as fires or heart failures, are so critical that the benefits of being constantly prepared outweigh the cost. For other events, such as flooding, snowstorms or blackouts, a shorter or longer delay may be acceptable depending on the conditions. It is simply prohibitively expensive to be ready to respond immediately to every possible event.

The threshold for responding is likewise important. If it is too low, an organisation will respond too often and too early and thereby waste resources. If it is too high, the organisation will respond too late or perhaps not at all. (Remember the story of the boy who cried wolf?) In short, how correctly does an organisation recognise that there is a need to respond, and how well do the prepared responses match the requirements of the situation?

The timing of the response is important in two ways. One is whether the response starts at the right time. Responding too late is a frequent worry, but it may also be problematic to respond too early – for instance close off an area before everyone inside has been evacuated. Another is the duration of the response, especially whether it can be sustained long enough. A search for survivors, for example, should neither end too early nor continue too long. Is there an organisational 'stop rule' to determine when normal operations can be resumed? The magnitude of the response must also be right. It is essential for an organisation to be able to provide a response of the necessary intensity or magnitude and sustain it as long as required. Bushfires, for instance, can go out of control when firefighters become exhausted or when materials (vehicles, firefighting chemicals) cannot be replenish fast enough. Vaccines can likewise become unavailable during an epidemic.

A final diagnostic question is about whether and how the required competence and resources are maintained and verified. This is relatively easy to determine in the case of material or physical resources. But it can be difficult for intangible resources, such as skills and competence. The competence of people is often a decisive factor, but how can it be ascertained that the required competence is available, and how can it be ensured that such competence is maintained over time?

Example: Competence assessment of air traffic controllers

In many professions, a regular check of competence is required. One of these is air traffic management, where international organisations responsible for the safety of air navigation, such as Eurocontrol, have issued guidelines for competence assessment. This includes four elements: (1) a continuous assessment by making observations of the assessment of air traffic controller (ATCO) during normal operational duties, (2) a dedicated practical assessment that is carried out annually, followed by (3) an oral and/or (4) a written examination. Pilots are also checked regularly in order to ensure that flights – and passengers – are safe. But for many other professions – doctors, for instance – that is not the case!

Questions addressing the potential to respond

Table 5.2 provides an example of a set of generic questions that can be used to assess an organisation's potential to respond.

A practical example of diagnostic questions for the potential to respond is shown in Table 5.3. This set of questions has been developed for use by a Canadian inner city emergency department (Hunte and Marsden, 2016).

Table 5.2 Examples of detailed issues relating to the potential to respond

Event list	Is there a prepared list of possible and potential events or conditions (internal or external) for which the organisation should be ready to respond?
Relevance of event list	Has the list been verified and/or is it revised on a regular basis?
Response set	Have responses been planned and prepared for every event in the list?
	Do people know what to do when one of these events occur?
Relevance of response set	Does the organisation check that the responses are adequate? How, and how often, is this done?
Response start and stop	Are the triggering criteria or threshold well defined? Are they relative or absolute?
	Are there clear criteria for ending the response and returning to a 'normal' state?
Activation and duration	Can an effective response be activated fast enough?
	Can it be sustained as long as needed?
Response capability	Are there sufficient support and resources to ensure response readiness (people, equipment, materials)?
Verification	Is the readiness to respond (response capability) adequately maintained?
	Is the readiness to respond verified regularly?

Table 5.3 An example of questions to assess the potential to respond

Question	Contents
1	We have a list of everyday and unexpected clinical, system and environmental events for which we prepare and routinely practice action plans.
2	We revisit and revise our list of events and action plans on a systematic basis.
3	We follow defined thresholds, actions and stopping rules to adapt/transform operations and proactively mobilise resources in order to maintain our capacity for response under conditions of increased volume and acuity.
4	We effectively team, communicate and work together within the department and with other departments and services.
5	We have organisational support and resources to maintain our capability to meet acuity and volume demands.
6	We link our local department adaptations to organisational and health system changes.

A comparison of Tables 5.2 and 5.3 shows that the analysis group at the emergency department has chosen a subset of the generic questions but also added some new questions, e.g., question #4. The specific form of the questions has also been changed from a neutral or technical frame, referring to the organisation as such, to a social frame that better reflects the intersubjective, interdependent and human nature of health care.

The potential to monitor

An organisation must be able to respond to events and changes, to threats and opportunities alike, but that is not sufficient to ensure its survival in the long run. The reason is simple: an organisation that does not keep an eye on what goes on in the operating environment will always be surprised when something happens. An organisation that constantly must respond to surprises will soon exhaust its resources and capabilities. Forewarned is forearmed, as the proverb goes. Responding to the unexpected is like fighting a bushfire or playing 'whack the mole'. It works as long as the fires are infrequent or as long as there is sufficient time between the appearances of the moles. But if the frequency increases so that the time between events is shorter than the time needed to orient and respond, and if furthermore the onset of events is unpredictable, then responses will be too late and ineffective.

The potential to respond is therefore of limited value unless there also is a potential to monitor. Conversely, an organisation that knows about or can guess what will happen before it happens will be ready to respond when the need arises. The purpose of monitoring is to keep a close watch on something or to sample data on a regular basis in order to understand both what goes on in an organisation and its operating environment (inside and outside the organisation's boundary) and in particular, whether something may develop in a way that will require a response.

The diagnostic questions related to the potential to monitor are about *what* is monitored (and measured) and *why*. These questions are at the heart of the indicator problem, something that has been discussed extensively for many years (see Chapter 4). One central issue is whether indicators are lagging or leading – whether they represent what *has* happened or what *will* happen. Others are whether an organisation uses the right indicators and makes the right measurements and whether the set of indicators is evaluated and revised on a regular basis.

If an organisation is satisfied that it uses the right indicators, the next questions are how they are used and how often they are looked at, checked or measured. An equally important question is whether the indicators are meaningful (valid) or just convenient? In practice – taken across all industries – indicators can be seen as representing a compromise between efficiency and thoroughness. In this context, the efficiency refers to how easy it is to measure or 'read' the current value of the indicator and how easy it is to compare it to other indicators or to commonly accepted references or criteria. Thoroughness refers to how meaningful an indicator is, both in terms of how valid it is relative to the condition or process it is supposed to represent and how well and how directly it supports decisions about corrective, supportive or remedial actions. A simple juxtaposition of the two criteria leads to the following categories of indicators:

- Indicators that are both easy to measure and meaningful. These are, of course, the ideal indicators, but practical examples are hard to find.
- Indicators that are easy to measure but are not very meaningful. The literature, and industrial practice, abounds with such indicators. Most indicators of safety and quality – at least for sociotechnical systems – are found here. 'Not everything that counts can be counted, and not everything that can be counted counts' as William Bruce Cameron once said. But few people have heeded that advice.
- Indicators that are difficult (or costly) to measure but are meaningful. These are the ones that 'count but cannot (easily) be counted'. They have been defined by a need to have specific information, even though there may be no obvious way of getting it. Indicators in this category are generally aggregated or calculated from other indicators that are easier to measure, and the aggregation reflects the meaning of the indicator. An example is the Consumer Price Index described in Chapter 4.
- Irrelevant indicators that are difficult (or costly) to measure and are not meaningful. Although this category obviously is of little practical interest, examples can nevertheless be found, not least in economics. Some indicators, such as the consumer sentiment index, that are indirect (calculated) may be used because of tradition, even though there is no evidence that they actually work. Another example is the level of safety culture.

If indicators are grouped according to these four categories, the result is a slanted ellipse as shown in Figure 5.2 with most indicators in the lower right part.

Other issues related to monitoring are the frequency of measurements and how quickly they are analysed and interpreted. The interpretation of the measurements must be fast enough to support timely interventions when called for. This may explain the widespread preference for simple indicators, particularly when they can be expressed quantitatively, as numbers. But the interpretation should be more than just a recognition that the value of an indicator at time n is higher or lower than a set point or a threshold or the value of the same indicator at time $n - 1$. Consider, for instance, the number of people that die in traffic accidents in a country. There is much focus on the absolute number, usually issued when a calendar year ends, but how can we know that two consecutive measurements are significantly different? More to the point, how can such a difference be interpreted so that it becomes the basis for effective remedial or corrective actions or interventions?

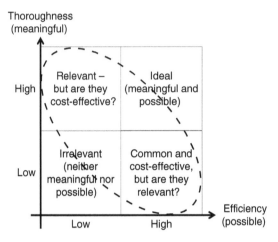

Figure 5.2 Distribution of indicators with regard to efficiency and thoroughness.

Example: hospital standardised mortality ratio (HSMR)

A hospital's standardised mortality ratio is the ratio between the observed number of deaths among patients admitted to a hospital during a period of time and the expected number of deaths (using a logistic regression model). It is used worldwide to assess and analyse mortality rates in order to identify where patient safety and the quality of care can be improved. It is attractive because it provides a single number, the HSMR, that intuitively seems to make sense. The HSMR has, however, been criticised as being potentially misleading because of low criterion validity and imprecise with uncertain stability over time. Its simplicity may therefore belie its value. The HSMR is an example of an indicator that is easy to measure but is not very meaningful.

Questions addressing the potential to monitor

Table 5.4 provides an example of a set of generic questions that can be used to assess an organisation's potential to monitor.

A practical example of diagnostic questions for the potential to monitor is shown in Table 5.5. This set of questions has been developed for use by the French National Railways (SNCF) (Rigaud et al., 2013).

In this case, the actual diagnostic questions are in most cases taken directly from the generic questions with additional questions added (e.g., question #7). The form of the diagnostic questions is a direct statement that in principle can be confirmed or denied. The actual answer categories are, however, more subtle as will be described in a following section.

Table 5.4 Examples of detailed issues relating to the potential to monitor

Indicator list	Does the organisation have a list of regularly used performance indicators?
Relevance	Is the list verified and/or revised on a regular basis?
Validity	Has the validity of indicators been established?
Delay	Is the delay in sampling indicators acceptable?
Sensitivity	Are the indicators sufficiently sensitive? Can they detect changes and developments early enough?
Frequency	Are the indicators measured or sampled with sufficient frequency? (Continuously, regularly, every now and then?)
Interpretability	Are the indicators/measurements directly meaningful or do they require some kind of analysis?
Organisational support	Is there a regular inspection scheme or schedule? Is it properly resourced? Are the results communicated to the right people and put into use?

Table 5.5 An example of questions to assess the potential to monitor

Question	Contents
1	The safety performance indicators are consistent with the organisation.
2	Indicators are revised regularly and properly.
3	The organisation uses leading indicators.
4	The organisation uses lagging indicators.
5	The leading indicators are valid.
6	The period covered by the lagging indicators is appropriate.
7	The type of measurements (qualitative or quantitative) is appropriate.
8	The frequency of measurements is appropriate.
9	The delay between the measurement and analysis of the results is acceptable.

The potential to learn

An organisation that has the potential to monitor, to keep track of what happens and the potential to respond when so required may function reasonably well as long as the conditions (demands, resources, operating environment, etc.) remain stable. If they change, the organisation must also change, which means that it must have the potential to learn. Learning is the active and deliberate modification of how an organisation copes with everyday situations, large and small. The essence of learning is therefore the ability to change how an organisation responds, monitors, anticipates – and how it learns.

One question is about the basis for learning. In relation to safety, the established wisdom is to learn from accidents. Accidents are unwanted occurrences, and it makes obvious sense to try to avoid these as far as possible. According to the *causality credo*, accidents have causes that can be found and – in principle – eliminated. In order to prevent accidents from happening again it is therefore necessary to learn from them, QED. To this conventional view of safety, a Safety-II perspective offers an alternative. If something that goes wrong has a cause, then something that goes well must have one too. Since many more things go well than go badly, it makes sense to try to learn something from them. Learning from failures alone is limiting, in addition to being rather expensive. An organisation should learn from everything that happens, from what goes well and from what goes badly – and from everything in between.

Learning can either be irregular or regular – or even continuous. Irregular learning takes place as a reaction to an unusual event, such as an accident. If something is sufficiently different from what it used to be or was expected to be (in either a negative or positive sense) it becomes a possible focus for learning. This type of learning is event-driven and reactive. It also assumes that when 'nothing' happens, as when there are no accidents or no positive surprises, then there is nothing to learn – and consequently nothing is learned. This way of reasoning is fundamentally wrong. Indeed, a great many things happen when 'nothing' happens, and an organisation can benefit hugely by learning from these.

Regular learning is based on performance patterns rather than unique events. This highlights the importance of how an organisation collects the data and information that are the basis for learning. Is it done as a natural part of everyday work (hence as an integral part of work), or separately – either at a specific time or by specialists? If specialists are in charge of learning, they have presumably been given the competence and the resources necessary to do their job. But if learning is a separate activity it easily becomes detached from everyday work. This introduces a time lag between when the 'lessons' are learned and when they can be turned into practice. Indeed, the further learning is removed from the actual experience, the larger the lag and the less the precision and the level of detail. This is illustrated by Figure 5.3, which shows the typical pathways of learning from events. The further a report has to travel through an organisation, the longer it takes for it to be noticed and for any feedback to be provided. Another consequence is that the information is gradually changed from raw data (a direct rendering of the actual events), through analysed data and aggregated data to performance indicators (such as scorecards) or longer term trends (statistics). Decisions made using high-level indicators clearly will be less concrete and operational than decisions taken on the spot.

Another diagnostic question concerns the resources allocated to learning. This is not unrelated to whether learning is direct and individual or mediated and organisational. Resources allocated to learning are often seen as a cost rather than an investment, not least during a period of relative 'calm', hence among the first to go when the financial conditions become more strained.

Finally, there is the question of how learning is implemented. Is the learning expressed as new rules and procedures, as modified training contents, as redesign

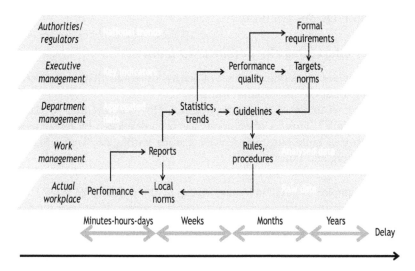

Figure 5.3 Delays in organisational learning.

of equipment, workplaces and organisational structures, or even as a limited re-organisation? Are the changes temporary or permanent, and are they themselves evaluated? How long is the effect of learning assumed to last? How are the lessons learned maintained, and how is it verified that something has actually been learned? Irregular learning, linked to accidents and disruptions, is difficult if not impossible to verify, since a similar situation may not happen again. Here, it is clearly an advantage to link regular learning to smaller evens or even everyday occurrences, since it is much easier to determine whether there has been any effect of learning.

Example: Alan Greenspan and the financial crisis

Learning, as a psychological phenomenon, is subject to several of the biases that shape human cognition. One of these is the confirmation bias, which leads people to prefer information that confirms their beliefs or assumptions. This is nicely illustrated by a statement in an interview with Alan Greenspan, former Chairman of the Federal Reserve of the United States, in *The Guardian* (October 24, 2008). Commenting on the financial crisis, Greenspan admitted that: 'Partially ... I made the mistake in presuming that the self-interest of organisations, specifically banks, is such that they were best capable of protecting shareholders and equity in the firms ... I discovered a flaw in the model that I perceived is the critical functioning structure that defines how the world works. I had been going for 40 years with considerable evidence that it was working exceptionally well'.

Greenspan's admission illustrates not only a failure to learn but also the absence of anticipation – or perhaps a 'mechanical' anticipation that is no more than a simple extrapolation from fixed and unchallenged premises. The very same development, the subprime bubble in 2007–2008, was correctly predicted by an outsider, Dr. Michael Burry.

Questions addressing the potential to learn

Table 5.6 provides an example of a set of generic questions that can be used to assess an organisation's potential to learn.

A practical example of diagnostic questions for the potential to learn is shown in Table 5.7. This set of questions has been developed by researchers at the University of Linköping, Sweden, for a study within Air Traffic Management (Ljungberg and Lundh, 2013).

Table 5.6 Examples of detailed issues relating to the potential to learn

Selection criteria	Does the organisation have a clear plan for which events to learn from (frequency, severity, value, etc.)?
Learning basis	Does the organisation try to learn from things that go well or does it only learn from failures?
Learning style	Is learning event driven (reactive) or continuous (scheduled)?
Categorisation	Are there any formal procedures for data collection, classification and analysis?
Responsibility	Is it clear who is responsible for learning? (Is it a common responsibility or assigned to specialists?)
Delay	Does learning function smoothly, or are there significant delays in the learning process?
Resources	Does the organisation provide adequate support for effective learning?
Implementation	How are 'lessons learned' implemented? (Regulations, procedures, training, instructions, redesign, reorganisation, etc.)

Table 5.7 An example of questions to assess the potential to learn

Question	Contents
1	It is clearly established what should be reported.
2	Submitted reports are being investigated sufficiently.
3	There are good responses/feedback on submitted reports.
4	The time from the submission of a report until a response is acceptable.
5	There are sufficient resources to write reports.
6	The employees are being motivated to write reports.
7	Lessons are learned from things that go right, as well as things that go wrong.
8	We meet with personnel from other units to learn from each other.

In this case, the diagnostic questions have been formulated to emphasise the individual view of the potential to respond. Questions #2, #3 and #4, for example, address the employees' concern for how learning can take place rather than organisational or systemic concerns. Question #8 is about 'indirect' learning from colleagues rather than 'direct' learning from collected data. The whole set of questions, as well as the questions for the three other potentials, reflect the specific nature of work in air traffic management and the concerns of ATCOs. In this case, many of the questions are also formulated in a way that allows binary answers, although this does not exclude that more graduated replies can be used as well.

The potential to anticipate

Nearly all organisations are concerned about their ability to cope with the fluctuations and disturbances that may affect everyday performance and try to strengthen this ability with how they manage their potentials for monitoring and responding. Many organisations worry about keeping up with changes in the operating environment, with new demands and with new constraints. Some organisations are also concerned about what the future may bring, not least if their vision includes a long-term perspective. But thinking about what may happen in the future, the potential to anticipate, is perhaps the least developed and least appreciated of the four potentials for resilient performance. However, resilient performance is not possible without the presence to some degree of all four potentials.

The potential to anticipate is more than an extension of the potential to monitor. Where the purpose of monitoring is to keep an eye on what happens, both inside and outside an organisation, the purpose of anticipation is to think or speculate about what may happen in the future. Where monitoring is about paying attention to what may become critical, anticipation is about imagining the potential. This is probably the main reason why relatively little has been done about anticipation, at least in the context of conventional safety management. Engineering and imagination are often seen as being incompatible, although nothing could be further from the truth.

Perhaps the most important condition for the potential to anticipate, and therefore the most important aspect to assess, is the existence of a corporate vision that accepts the need to spend time and effort to think about the future. Many organisations do that when they try to predict their 'market' – whether it is the future customers, the future patients, the future regulations and restrictions (legislation), etc. But the potential to anticipate goes beyond that. It is not just having an idea of what may happen, but also an idea about what an organisation would like to achieve or accomplish, or what (or where) it would like to be, in the long term. Closely linked is the willingness to take risks. Without it, there will be little gained from anticipation. Risk aversion is the enemy of anticipation.

An important part of anticipation is the time horizon, how far an organisation is willing to look into the future. For some organisations, the answer is given by the nature of their activity. For instance, it is necessary to look far ahead when you plan to build a nuclear power plant, a motorway bridge, a windmill farm or a hospital, because it is a huge investment, because the expected lifetime is long and because it is not something that you can simply leave or quickly dismantle. For other organisations, it is a matter of ambition – and of tolerance of uncertainty.

If an organisation engages in anticipation, a further question is how the organisation thinks about the future. In other words, what is its model of the world and of the events that happen in it? What does it assume about the dynamics of the events that may affect it? Where do possible threats and opportunities come from? Is anticipation based on an articulated model or is it based on hunches, feelings or intuitions? Related to that is the question of whether there is a clearly formulated strategy – and whether is it commonly known.

Two final diagnostic questions are when does thinking about the future take place and who does this thinking? Is it a regular or irregular concern, something that is driven by the corporate vision, by the example set by a charismatic leader, or is it something that happens in response to an overwhelming change or a disaster? Is anticipation outsourced to a consultancy or a think tank or is it something that is done inside the organisation – and then by whom?

Example: Turing Pharmaceuticals

The pricing policy of Turing Pharmaceuticals illustrates the consequences of not anticipating the response from the public and the market. In September 2015, the company increased the price of one of its products, Daraprim for the treatment of a parasitic infection, by 5.555% from $13.50 to $750 a pill. The company argued that the profit from sales would be used to develop new therapeutics that hopefully would help eradicate the disease. Not unsurprisingly – except to the company – there was widespread public outrage as well as questions from two U.S. congressmen. The company's response was to hire four lobbyists and a crisis public relations firm to help explain the pricing decision.

On December 17, 2015, Turing's CEO was arrested by the FBI on (unrelated) charges of running a Ponzi scheme at his former company. The company did not reduce the price of the drug, but other companies stepped in with offers of lower-cost alternatives.

Questions addressing the potential to anticipate

Finally, Table 5.8 provides an example of a set of generic questions that can be used to assess an organisation's potential to anticipate.

A practical example of diagnostic questions for the potential to anticipate is shown in Table 5.9. This set of questions has been developed by the Australian Radiation Protection and Nuclear Safety Agency as part of their Holistic Safety Guidelines (ARPANSA, 2012).

Table 5.8 Examples of detailed issues relating to the potential to anticipate

Corporate culture	Does the corporate culture encourage thinking about the future?
Acceptability of uncertainty	Is there a policy for when risks/opportunities are considered acceptable or unacceptable?
Time horizon	Is the time horizon of the organisation appropriate for the kind of activity it does?
Frequency	How often are future threats and opportunities assessed?
Model	Does the organisation have a recognisable and articulated model of the future?
Strategy	Does the organisation have a clear strategic vision? Is it shared'?
Expertise	What kind of expertise is used to look into the future (in-house, outsourced)?
Communication	Are the expectations about the future known throughout the organisation?

Table 5.9 An example of questions to assess the potential to anticipate

Question	Contents
I	What systems are in place to look into the future at potential safety and security related weaknesses and threats?
2	Do these systems or people who make these forecasts have sufficient expertise, capability and resources to make accurate and relevant forecasts?
3	How far ahead and how frequently are these forecasts made?
4	What criteria are used to determine the scope and depth of these forecast analyses?
5	What systems are in place to ensure it is easy, straightforward and welcoming for staff to raise any issues related to potential or anticipated safety and security-related weaknesses and threats? How are these staff contributions taken into account?
6	What systems are in place to ensure that forecast information is communicated to relevant parts of the organisation? Is this information adequately conveyed and shared with relevant staff and departments/processes?
7	What systems are in place to develop and maintain staff skills and competencies to adequately anticipate future safety and security weaknesses and threats?
8	What systems are in place (where appropriate) to ensure control measures are developed and implemented to address issues raised in forecast analyses? Are staff adequately consulted in this development and implementation process?

In this example, the diagnostic questions represent the concerns of a regulator rather than of the organisation itself, although there naturally is an overlap between the two. Each question thus begins with the same words, 'What systems are in place ...'. The questions are directed at the management of the organisation, although that does not preclude that they can be answered by others as well. The management focus is legitimate for a regulatory body since this is where potential remedial actions must be aimed. Some of the questions require rather detailed answers while for others replies can be given by pre-defined categories. The type of answers sought must clearly be suited to the purpose of the assessment, which in this case is to evaluate the management of the organisation rather than manage the organisation directly.

Proxy measures

An assessment or measurement of the potentials for resilient performance cannot be based on samples of distinct events or performance fragments. It must refer to the characteristics of performance as it unfolds over weeks or months. An organisation would ideally have to be followed for a period of time to determine whether or not the performance was resilient. Since this is rarely feasible, some sort of proxy measure must therefore be found that satisfies demands to efficiency as well as to thoroughness.

Resilience is defined as an expression of how people, alone or together, cope with everyday situations – large and small – by adjusting their performance to the actual operating conditions. Resilience is therefore associated with individual and organisational performance (processes) rather than outcomes (products). But it is notoriously difficult to measure individual and organisational performance directly because it is intractable and because it is underspecified. A solution is provide by the four potentials that are the basis for resilient performance. This does not mean that each potential should be treated as if it were a simple, uniform quality represented by a single measure. The proxy measures should instead be based on the details of each potential as explained in this chapter. The advantage is that such proxy measures change more slowly than the actual process (performance). It will also be easier to change and manage them via their constituent functions or processes – while considering the dependencies.

Formulating the diagnostic questions

The assessment of the potentials should be detailed enough to make it possible to manage an organisation. Replies to the diagnostic questions should therefore be graduated rather than binary (such as 'yes' versus 'no', 'agree' versus 'disagree', etc.), for instance by using a Likert-type scale. The four examples of actual diagnostic questions show different ways of formulating the questions and of recording the answers.

- Table 5.3 (Canadian inner city emergency department) formulated the questions so that the respondents could express their level of agreement – or disagreement. In principle, this can be as a binary answer – agree or disagree – but a more common and more useful version is to allow the respondents to specify their level of agreement or disagreement on a symmetric scale. A typical five-level Likert-type scale could use the following answer categories: 'strongly disagree' – 'disagree' – 'neither agree nor disagree' – 'agree' – 'strongly agree'.
- Table 5.5 (SNCF) used the same approach, but in this case the questions were accompanied by four answer categories: 'satisfactory' (*satisfaisant*) – 'acceptable' (*acceptable*) – 'medium' (*moyen*) – 'insufficient' – (*insuffisant*).
- Table 5.7 (LiU) also used a Likert-type scale, this time with the following five answer categories: *Excellent* (the organisation on the whole exceed the criteria addressed by the specific question), *Satisfactory* (the organisation fully meets all reasonable criteria addressed by the specific question), *Acceptable* (the organisation meets the nominal criteria addressed by the specific question), *Unacceptable* (the organisation does not meet the nominal criteria addressed by the specific question) and *Deficient* (there is insufficient capability to meet the criteria addressed by the specific question).
- Table 5.9 (ARPANSA) used a different style, where the questions were more open. Some required a detailed answer rather than a level of agreement/disagreement. One example is the question: 'What systems are in place to develop and maintain staff skills and competencies to adequately anticipate future safety and security weaknesses and threats?' Some asked for a level of agreement/disagreement like the following: 'Do these systems or people who make these forecasts have sufficient expertise, capability and resources to make accurate and relevant forecasts?'

Since the RAG is intended to be administered repeatedly over a period of time, it may be useful for the diagnostic questions to be administered in some standard form, for instance, electronically via email or a website, thus reducing the need for face-to-face meetings or interviews. It goes without saying that the actual formulation and presentation of the questions should pay attention to the best practices of human factors and social sciences.

How to present the results of an assessment

A distinct advantage of using a Likert-type scale is that the results can be shown directly in a table or in a variety of graphical renderings, such as bar charts, diverging stacked bar charts, multiple pie charts, square pie charts, etc. The choice of an effective way to show the results should keep in mind that the assessments are not one-off measurements but repeated measurements that are intended to support managing a process or a development. It is therefore useful if the results from one assessment can be easily compared with the results from another. Such a comparison can show the magnitude and direction of any changes that may have occurred.

A radar chart or star plot is an effective way to present the results of an assessment. A radar chart uses a number of equiangular spokes where each spoke represents one of the questions and the length of a spoke is proportional to how the question was rated by the respondents on a Likert-type scale. The interest here is, however, not to determine the distribution of answers, which is how a Likert-type scale normally is used, but rather to determine the shared view, expressed by, e.g., the mean or the median of the replies. The result is a star-like polygon, which provides a clear signature of how the answers are distributed with regard to the particular potential.

Figure 5.4 shows what the ratings for the potential to respond could look like for an (unspecified) organisation. (The questions refer to the generic set of questions shown in Table 5.2 and use the standard five-level Likert-type scale for answers. The example is fictitious.) For this organisation, it is assumed that assessments take place in 4-month interval. If a value of '1' corresponds to a low score on the Likert-type scale and a value of '5' corresponds to a high, the irregular shape of the polygon for Month 4 in Figure 5.4 clearly shows that not all functions (as addressed by the questions) are seen by the respondents as being satisfactory or even adequate. The response capability (in terms of start/stop, speed and duration) and the verification of the responses are, for instance, rated as unsatisfactory. But it should also be noted that both the event list and the response set are seen as adequate and that there are sufficient resources set aside to allow the organisation to respond when needed. The organisation should use

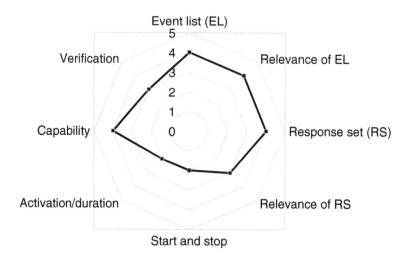

Figure 5.4 Assessment of the potential to respond (Month 4).

this assessment, seen in the context of the assessment of the three other potentials, as a basis for planning both how to improve the functions that were rated as unacceptable and how to sustain the functions that were seen as adequate. The organisation should look at weaknesses and strengths together and strive to maintain a proper balance of the functions that make up the four potentials, as well as the four potentials themselves. It would be a misuse of the RAG, if only 'low' scores were addressed and even worse if they were addressed in isolation.

Developing the potential to respond

Looking at the responses shown in Figure 5.4, the answers to four of the eight questions obviously leave something to be desired while the answers to the remaining four questions are acceptable. The situation can be described as follows.

- The relevance of the response set (RS), the prepared list of events or conditions for which the organisation should be ready to respond, is unacceptable, corresponding to a score of '3'.
- Neither the triggering criteria for starting a response nor the criteria for ending a response are clear, receiving a score of '2'.
- According to the assessment answers, it takes too long to activate an effective response, and it is uncertain whether it can be sustained as long as needed (score of '2').
- The organisation's readiness to respond is not verified regularly (score of '3').

Because of the diagnostic and formative nature of the questions, the answers provide a useful starting point for thinking about how the situation could be improved and for proposing concrete remedial activities. Continuing with the same example, the recommendations could, for instance, be as follows:

- Review the responses that have been prepared for each event in the Event List. Such a review should be made on a regular basis and not just after a failure to respond. Even if there have been no problematic situations, this may be because the organisation has been 'lucky' rather than because it has been completely prepared.
- The assessment shows that the triggering (start) and stop criteria are seen as insufficiently clear, at least for a subset of the responses. It would therefore be reasonable to go through each planned response to look at whether the conditions are well defined and clearly described or whether a revision is needed. Ditto for the stop criteria, where it additionally should be considered when and how the organisation will be able to resume is primary functions. Failure to do so is tantamount to organisational complacency.
- If the delay between the occurrence of the triggering conditions and the execution of the response is too large, the organisation should consider whether its preparations are adequate and specifically whether the

resources have been correctly allocated. This may lead to a reassessment of the organisation's goals and priorities. Some events may require an immediate response and some may be able to tolerate a delay. But deciding which is which requires careful consideration, both in terms of how well it has worked in practice (experience or lessons learned) and in terms of how (limited) resources should be allocated. Similar considerations must be made for the capacity to sustain a response long enough – including a decision on what 'long enough' means in practice.

• Finally, the organisation should look at how it establishes that the readiness to respond actually is in place. Does it, for instance, rely on safety cases that are made once but never again? Does it take into account that requirements to resources (material, competence) may change over time or that some resources may deteriorate? Does it rely on checklists or negative reporting, or does it actively seek the viewpoints of the people who are directly responsible? Is the readiness to respond checked regularly or only in the aftermath of a serious adverse event?

The other four questions received ratings that showed the corresponding functions to be acceptable. It may therefore seem less urgent to address them. But any organisation should remember that safety is a 'dynamic event'. This means that the acceptable outcome of the assessment does not happen by itself, but only because of specific efforts that must be sustained. It is just as important to pay attention to and support what an organisation does well as it is to improve what is unsatisfactory.

To continue the example, assume that the same organisation carried out an assessment 4 months later, producing the outcomes as shown in Figure 5.5.

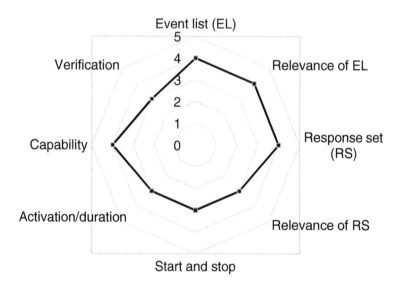

Figure 5.5 Assessment of the potential to respond (Month 8).

The differences between the two assessments are easy to see and can be used to determine both whether the organisation develops in the right direction and where specific interventions should focus. The second radar chart shows where improvements have been made and where improvements are still needed. If we look at Figure 5.5, the following changes are noticeable.

- The efforts used to revise the response set have not succeeded. This may be because it takes longer to effectuate the change or because it takes more than 4 months for the results to be visible.
- The assessment of the triggering criteria for starting a response and the criteria for ending a response is better, with a score of '3' instead of '2'. There is, however, still room for improvement, and the organisation should continue its efforts.
- There has also been some improvement with regard to how fast an effective response can be started and/or how long it can be sustained. A score of '3' may, however, not be fully satisfactory, and the organisation should continue its effort.
- Finally, there has not been the expected improvement in how the readiness to respond is verified. The reasons may be the same as for the first point (appropriateness of the response set).

A lack of expected improvement may be because the chosen intervention was not the correct one or because it was based on too simple an understanding of how the organisation works. In the latter case, additional efforts may be needed to develop a reasonably accurate understanding of how the organisation works. How this can be done will be discussed in the following chapter.

The radar chart in Figure 5.5 also shows, reassuringly, that the organisation has succeeded in maintaining the other four functions.

Diagnostic and formative questions

As pointed out at the beginning of this chapter, the questions in the RAG must be diagnostic as well as formative. They must be diagnostic in the sense that they focus on specific and concrete issues that are important for a potential or for a facet of a potential. And they must be formative in the sense that the answers can serve as a basis for proposing specific activities or interventions that will improve the potentials. The diagnostic and formative sides of the questions can be illustrated by looking at four examples.

The potential to respond

One of the questions used to address a subfunction of the potential to respond (Table 5.3) was 'We revisit and revise our list of events and action plans on a systematic basis'. The question is diagnostic because it looks at whether an organisation systematically revisits the conditions or events for which it maintains a readiness to respond. The question is not just whether an organisation has a readiness to respond to certain events or conditions, but also whether it systematically considers the relevance of this list of events and conditions. The question is formative because the answer clearly indicates what should be done. If the list is adequate, then an organisation should take steps to ensure that it remains so. And if it is not adequate – or seen as adequate – then something must be done to overcome that, for instance, introducing and supporting a scheme to revise the list, to ensure qualified staff, to give appropriate priority – and time, manpower and resources – to improve the list.

The potential to monitor

One of the questions used to address a sub-function of the potential to monitor (Table 5.5) was 'The organisation uses leading indicators'. The question is diagnostic because it inquires whether an organisation relies exclusively on lagging indicators, or whether it combines lagging and leading indicators. This might be pursued further by taking a closer look at the indicators that are used for instance with regard to their origin or source, their documented practical value, etc. The question is also formative because a negative answer, meaning that an organisation relies exclusively on lagging indicators, can be the starting point for considering how monitoring can be improved. Such improvements must obviously be tailored to the needs of the organisation. Yet every organisation must to some extent keep an eye on trends, on how it is likely to develop in the near and mid-term. Doing that well may not always be easy, but knowing that it is not done at all is a clear warning that something needs to change.

The potential to learn

One of the questions used to address a subfunction of the potential to learn (Table 5.7) was 'The time from the submission of a report until a response is acceptable'. The question is diagnostic because it queries an important condition for learning, namely, that information – or feedback – is made available while the memory of the event is still fresh and before working conditions have changed too much. If the delay is too long, months or years rather than days or weeks, cf., Figure 5.3, the information may no longer be relevant. Staff may have changed, the nature of work may have changed or the operating environment may have changed. If either is the case, the response may no longer make sense to the people who receive it, and therefore simply be filed in a ring binder or added to a database. In neither case will there by any useful learning. The question is also formative because a negative answer points to remedial actions with a concrete purpose: reduce the time it takes to analyse reports and formulate responses. The solution is, of course, not just to increase the pace of work or demand of people that they reduce the turn-around time. The remedial actions must instead be based on a thorough understanding of how reports are processed, the bureaucratic or administrative structures, the priorities of the organisation, the availability of competent personnel, etc. The focus could, for instance, be on how reports are written and submitted, how the information is processed and analysed or how analyses are documented and reported.

The potential to anticipate

One of the questions used to address a subfunction of the potential to anticipate (Table 5.9) was 'What systems are in place to ensure that forecast information is communicated to relevant parts of the organisation?' The question is diagnostic because it looks at how well the internal communication of an organisation works, especially when it comes to 'nontechnical' information such as forecasts and expectations for an organisation's future. Details to be investigated could be the form of the communication such as whether it is push or pull, what the target audience (recipients) is, whether the forecasts are described in a way that is interesting and understandable for the target audience, how often it happens, whether it is given special attention, etc. The question is also formative since it can easily give rise to concrete suggestions for how to improve the communication. In this case, as in all other cases, such suggestions should themselves be considered carefully in terms of both intended and unintended effects before they are implemented.

How to use the RAG to manage the potentials for resilient performance

The sets of diagnostic RAG questions that are developed for a specific use of the RAG should be formulated so that they can easily be assessed. They should therefore refer to concrete relations or characteristics of an organisation's performance, to something that the respondents have experience with or something that is described in the organisation's documentation. This has the added benefit that the questions themselves can serve as a basis for interventions to improve the resilient potentials.

The purpose of the RAG is to provide a well-defined characterisation (or profile) of an organisation that can be used to manage and develop an organisation's potentials for resilient performance. The RAG is intended to be applied regularly in order to keep track of how an organisation changes and develops. The RAG can thus be used to monitor organisational changes, which is a prerequisite for being able to manage performance effectively.

The RAG describes the potentials in terms of the constituent functions rather than as separate overall abilities. For each constituent function, the possible means to make the desired change can be developed and evaluated in terms of cost, specificity, risk, etc. Since the constituent functions of the four potentials may differ significantly among organisations, there is no standard or generic solution. But once the functions have been analysed and the goals defined, a variety of well-known approaches can be applied.

This can be illustrated by using the (fictive) case shown in Figure 5.4. Consider, for instance, the function *response start and stop* (Table 5.2), which looked at whether the triggering criteria or thresholds were well defined, whether they were relative or absolute, and whether there were clear criteria for ending the response and returning to a 'normal' state. In the assessment shown for Month 4 (Figure 5.4), the answer was 'unacceptable'. Assume that further probing revealed that the problem was due to the lack of a clear criterion for when a 'normal' state had been reestablished. The management could then use that as a basis for specific improvements or interventions. The result would then be as shown in the assessment at Month 8 (Figure 5.5), where the answer to the same question now is 'acceptable'. Although the example is fictive, it suggests how the RAG can be used in practice.

In terms of managing changes, the RAG is useful because it produces a profile of the current situation. For such a profile, it is clearly desirable that the number of details, the degree of resolution, is as high as possible. That is why using only the four potentials or – even worse – using a monolithic construct such as the level of safety culture or the level of resilience is too rough and imprecise. Neither is a comparison against other organisations useful for the purpose of managing change. Being better than another organisation – or being the industry leader – should never be a goal in itself, and can therefore not be the real target. But changing an organisation's performance in an intended direction is and should be a target. The assessment must always be specific, which means that it must be tailored to a known organisation.

To summarise, there are five important points to remember when using the RAG.

- Develop a tailored set of diagnostic and formative questions for the organisation. This must be based on a substantial experience with how the organisation functions. Such experience can be obtained by using a focus group, a discussion group or something similar. When developing the diagnostic questions it is also important to agree on the answer categories. If there are already known issues or problems with how an organisation functions, an attempt should be made to include these under one of the four potentials.
- Develop a description or model of the mutual dependencies among the four potentials. This is necessary both to interpret the data collected by the RAG and to develop effective responses (remedial actions). Such a representation or model must be specific to the organisation being managed. While it is possible to suggest a general model as a starting point (as it will be described in Chapter 6), the model must be tailored to the actual organisation. More importantly, the model must express not only the couplings between the four potentials but also how the four potentials depend on the more detailed subfunctions that are addressed by the diagnostic questions.

- Apply the RAG to the respondents, i.e., to (a subset of) the people who actually do the work. Collate the results and present them to stakeholders and respondents – and to the organisation as a whole. Discuss the conclusions and where changes are needed. Design remedial actions to bring about the changes.
- Try to work with a stable set of respondents so that they are the same people who provide the answers for a series of assessments. The purpose of getting answers to the RAG is not to produce a distribution of attitudes and views among the respondents, but rather to recognise the common view they represent.
- Be prepared to use the RAG in the long term and to make repeated assessments rather than a single measurement or rating. Managing and changing how an organisation performs, regardless of the type of performance or the criterion, must be done continuously over an extended period of time.

Finally, it is essential to acknowledge – and to keep in mind – that the four potentials are interdependent (Chapter 6), just as the functions relating to each potential are (Chapter 7).

RAG – towards a model of resilient performance

The Resilience Assessment Grid (RAG) provides a basis from which specific sets of diagnostic questions can be developed, but they are not intended to be used off the shelf. Since the diagnostic questions must be applied to a specific organisation, some clarification and reformulation will always be necessary.

Chapter 5 outlined the principles for how assessments can be rated and how the results can be presented. The radar chart is a compact representation of how the various items are rated and shows how well an organisation seems to be doing on each of the four main potentials at a given point in time. The RAG may with some justification be seen as a snapshot of how well the organisation performs – unlike an accident investigation, which is a snapshot of how an organisation has failed.

It has been mentioned several times that the four potentials for resilient performance depend on each other. This is clear both from the definition of the four potentials (Chapter 4) and from the detailed descriptions (Chapter 5). The ways in which the four potentials depend on each other have consequences for how they can be managed. It is clearly neither advisable nor in practice possible to manage them independently of each other. In this respect, the use of the RAG and the very idea of resilient performance differ from many other approaches to system safety – not least those that ascribe an all-dominating position to safety culture.

In order effectively to manage an organisation and to improve its performance with regard to some criterion (or criteria) – be it resilience, safety, quality or something else – it is necessary to understand how the four potentials depend on each other. An initial description can be derived from how they were defined and described. But the understanding should preferably be given a more practical or operational form, so that it can serve as a point of reference both for recommending specific changes and for analysing their consequences. In this way, the understanding serves as a model of how an organisation functions, not with regard to its primary activity (e.g., production, distribution or transportation) but with regard to how it controls what it does and how it manages specific aspects such as safety, productivity or quality.

Structural models of organisations

It is not difficult to find models of organisations either as representations of organisational structures or as descriptions of organisational flows (information or control, for instance). The first type, which represents the architecture of a formal organisation, is represented by the stereotypical hierarchical model with the blunt end (directors, generals) at the top and the sharp end (workers, soldiers) at the bottom. A generic model of a hierarchical organisation is shown in Figure 6.1. In this model, the components (the boxes) represent organisational roles, departments or units. The relationship is typically the hierarchy of superordinate-subordinate relations – or of who controls whom, for instance, that the Production Department controls the Factory Manager, while it in turn is controlled by the Managing Director.

Models of the second type, which represents how control (influence, leadership, information, etc.) flows through an organisation are all based on the standard input-throughput-output format, possibly with a feedback loop or two added. In the example shown in Figure 6.2, the input is represented by the 'external environment' and the output by the 'individual and organisational performance'. The model components (the boxes) are no longer organisational departments or units but instead represent factors that are accepted as being essential for the organisation's performance. The model illustrates that these factors affect each other but does not describe how the organisation functions. The model describes 'individual and organisational performance', but it says nothing about what that performance is, i.e., what the primary activity of the organisation is.

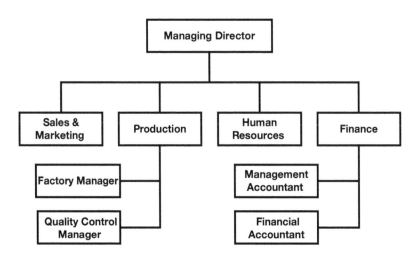

Figure 6.1 A generic, hierarchical organisation model.

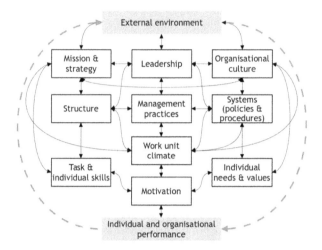

Figure 6.2 A causal model of organisational performance and change.
Source: (Burke & Litwin, 1992).

A third type is the strategy map, an integral part of the strategic management system, known as balanced scorecards (Kaplan and Norton, 1992). A strategy map is a diagram that describes how an organisation can create value by connecting strategic objectives in explicit cause-effect relationships with each other in the four BSC objectives (financial, customer, learning and growth). The strategy map consists of a number of perspectives, which in Figure 6.3 are called 'Financial', 'Customer', 'Internal Business Processes' and 'Learning and Growth'. Each perspective comprises a number of specific objectives; 'Customer', for instance, has two objectives called 'Exceed customer expectations' and 'Inspire loyalty'. The objectives can be linked both vertically by strategic themes that span the perspectives and horizontally across the perspectives. The strategic themes represent an organisation's hypotheses about how the strategy will produce the desired outcomes.

The main shortcoming of traditional structural models is that they describe a fixed structure or organisation of something, often called factors, which in Figure 6.2 are as diverse as 'mission and strategy', 'management practices' and 'motivation'. Structural models identify the main parts or components of an organisation and show how the various components are linked or connected. In Figure 6.1, 'Finance' is connected to 'Management Accountant' and 'Financial Accountant', meaning that the former controls – or is in charge of – the two latter. In Figure 6.2, 'Tasks & individual skills' is connected to 'Mission & strategy', 'Structure' and 'Motivation'. In Figure 6.3, 'Create high-quality products'

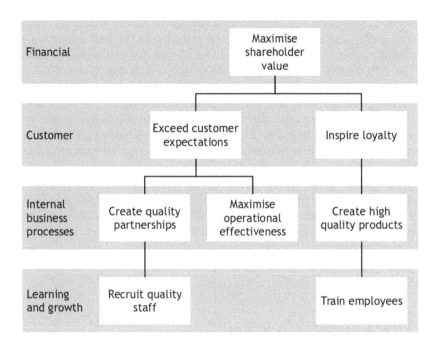

Figure 6.3 **A generic BSC strategy map.**

is linked to both 'Inspire loyalty' and 'Train employees'. But the connections, the 'arrows', merely signify that two components depend on each other in one way or another without providing any kind of detail about what the possible influences or connections are. (The situation is not improved by the double arrows. While they indicate that the influence goes in both directions, does this mean that the influence in one direction is the inverse of the influence in the opposite direction?)

Functional models of how an organisation works

The alternative is to develop a model that describes functions rather than parts or components (organisational structures) and that makes clear both which functions depend on each other and what form the dependencies take. In the case of the RAG, the natural starting point is the four potentials for resilient performance. Since a potential represents an organisation's ability to do something, it is quite natural to think of a potential as representing a function; the potential to respond can, for instance, be expressed as a function of responding and formally be represented by a function called <To respond>. The same goes for the three other potentials.

The Functional Resonance Analysis Method (FRAM; Hollnagel, 2012) provides a systematic way to produce a functional model of how an organisation works – in this case, starting by considering the four potentials as representing four functions. (A primer for the FRAM can be found at the end of this book.) The basic principle is that a function is described by what it *does* rather than by what it *is*; it is described by what the outcome or output is and by what is needed to produce the output. The description of a function therefore typically includes its Inputs and Outputs but may also include four other aspects called Preconditions, Resources, Control and Time. The Preconditions represent something that must be true or verified before a function can begin, the Resources represent that which is needed or consumed while a function is carried out, the Controls represent that which supervises or regulates a function while it is carried out and Time represents the various ways in which time and temporal conditions can affect how a function is carried out. (The following description follows the FRAM syntax so that function names are written with a capital in angled brackets, e.g., <Respond>, while the aspects are written in square brackets, e.g., [Priority areas]).

The four potentials as functions

The function <Respond> – or more correctly <To respond> – represents what an organisation does in a given situation. The Output of the function is therefore simply the [Responses] or the actions that an organisation takes to regain control of the situation. The function <Respond> first needs some inputs, which represent the conditions or triggers for the response. In this example, the two main Inputs are the [Interruptions] that come from the external processes and the [Alarms] that come from the function <Monitor>. (A further possible input, although not shown in Figure 6.4, could be [New demands], which could be the Output from a function called, for instance, <Manage production>.) The external processes represent an organisation's primary activities as well as what happens in the organisation's operating environment. In the model, this can, for instance, be subsumed under a single function called <Perform primary functions>.

Using the format prescribed by the FRAM, the function <To respond> may look as follows (Table 6.1).

The function <Monitor> represents the activities by which an organisation maintains its understanding of what is happening – in the operating environment as well as in the organisation itself. One important Output from <Monitor> is the [Alarms] that, as described above, serve as an Input to the function <Respond>. Another possible Output could be alerts, but these need not be considered in the first iteration of the model. The Inputs to the function <Monitor> are the relevant indicators and measurements from the primary functions on which the organisation keeps an eye. One example could be the [Process trends] that characterise how the external processes seem to develop, represented by the function

Table 6.1 A FRAM description of the potential to respond

Name of function	Respond
Description	An organisation's ability to respond to what happens or may happen

Aspect	Description of aspect
Input	Alarms
	Interruptions
Output	Responses
Preconditions	(none)
Resource	(none)
Control	(none)
Time	(none)

Table 6.2 A FRAM description of the potential to monitor

Name of function	Monitor
Description	An organisation's ability to monitor what happens around it and inside it

Aspect	Description of aspect
Input	Process trends
Output	Alarms
Preconditions	(none)
Resource	(none)
Control	Priority areas
	Lessons learned
Time	Sampling frequency

<Perform primary functions>. Monitoring must also be planned and managed, it must be controlled; a Control for <Monitor> could be the [Priority areas] that would be a sensible Output from the function <Anticipate>. Finally, <Monitor> may have a temporal aspect, Time, called [Sampling frequency], which specifies how often indicators should be read. The [Sampling frequency] could reasonably be an Output from the function <Learn>.

Using the format prescribed by the FRAM, the function <To monitor> may look as follows (Table 6.2).

The function <Learn> represents what an organisation does to gather and make use of the available experience. Although the potential to learn clearly involves a number of things, it is reasonable for the first iteration of the model to represent the potential by a single function. A main Input to <Learn> is, of course, the [Responses] that are the Output from <Respond>. An organisation will typically evaluate the [Responses] to determine how successful they were; although this is mostly done when the responses do not result in the expected outcomes, it is just as important to do it when the responses go well. One Output from this evaluation can simply be called [Lessons learned]. The [Lessons learned] may in turn be an essential Input for <Anticipate> as well as a possible Control for <Monitor>. Another Output from <Learn> can be the [Sampling frequency] that is a Time input to <Monitor>.

Using the format prescribed by the FRAM, the function <To learn> may look as follows (Table 6.3).

Finally, the function <Anticipate> represents the ways in which an organisation considers what may happen in the future. One important Input to that is obviously the [Lessons learned] since these summarise what went well and what did not. One result or Output from <Anticipate> could be the [Priority areas] that already have been mentioned and which in turn can be used to direct or control the function <Monitor>.

Using the format prescribed by the FRAM, the function <To anticipate> may look as follows (Table 6.4).

Table 6.3 A FRAM description of the potential to learn

Name of function	Learn
Description	An organisation's ability to learn

Aspect	Description of aspect
Input	Responses
Output	Lessons learned
	Sampling frequency
Preconditions	(none)
Resource	(none)
Control	(none)
Time	Latency of results

Table 6.4 A FRAM description of the potential to anticipate

Name of function	Anticipate
Description	An organisation's ability to anticipate what could happen in the future

Aspect	Description of aspect
Input	Lessons learned
Output	Priority areas
Preconditions	(none)
Resource	(none)
Control	(none)
Time	(none)

In addition to the four basic functions <Respond>, <Monitor>, <Learn> and <Anticipate>, the first iteration also introduced two other functions called <Perform primary functions> and <Regain control>, respectively. Both make good sense. One background function, <Perform primary functions>, represents what an organisation actually does or produces, whether it is materials or services. The other background function, <Regain control>, represents the activities defined by the responses that serve to bring the situation under control. The rationale for the two new functions is that every aspect that is defined for a function, in this case the four basic functions, must either come from somewhere or go to somewhere. Something cannot come from nothing, and something cannot disappear into nothing. The FRAM specifies that an aspect always must relate to at least two functions. There must be at least one function for which the aspect is defined as an Output, and there must be at least one other function where the aspect is used as either Input, Precondition, Resource, Control or Time.

In the FRAM terminology, the two new functions are initially classified as background functions. They can be described as follows (Tables 6.5 and 6.6):

Table 6.5 A FRAM description of <Perform primary function>

Name of function	Perform primary function
Description	To meet a demand is the generic background function that may create a need to respond – in addition to the normal production.

Aspect	Description of aspect
Input	(none)
Output	Interruptions
	Process trends
Preconditions	(none)
Resource	(none)
Control	(none)
Time	(none)

Table 6.6 A FRAM description of <Regain control>

Name of function	Regain control
Description	The response reestablishes control of the system (internal/external developments)

Aspect	Description of aspect
Input	Responses
Output	(none)
Preconditions	(none)
Resource	(none)
Control	(none)
Time	(none)

With the introduction of the two new functions, the first iteration of the model is completed. The outcome of this deliberately simple analysis of the four potentials as functions is shown graphically in Figure 6.4. (In the FRAM, functions are represented by hexagons.) The graphical rendering of the basic FRAM model represents the fundamental ways in which the four functions (representing the four potentials for resilience) depend on each other and therefore illustrates the first step in developing an understanding of how an organisation works. Even this first version of the model makes it clear that it is inadvisable to manage the four potentials separately from each other. Any suggestion for change or improvement must try to consider what a change in one function or potential would mean for the others. As an example, consider the potential to monitor, corresponding to the function <Monitor>. Assume that the answers to one or more of the (generic) questions in Table 5.4 show that there is a need to improve how monitoring is done. This might be achieved by changing the way monitoring is carried out in terms of frequency or sensitivity, for instance, but it might also be achieved by revising the list of indicators on which monitoring keeps an eye. The list is essentially the [Priority areas] that are Output from <Anticipate>, which means that an improvement of monitoring should include a consideration of how anticipation takes place.

In Figure 6.4, the functions of a FRAM model are represented by the four hexagons. (Background functions are in this rendering shown as shaded boxes.) Each function is described by what it does – corresponding to the name of the function – as well as by six aspects that may affect how the function works. It is, however, not necessary to describe all aspects for every function. The six aspects are represented by the letters at the six vertices of the hexagon: Input (I),

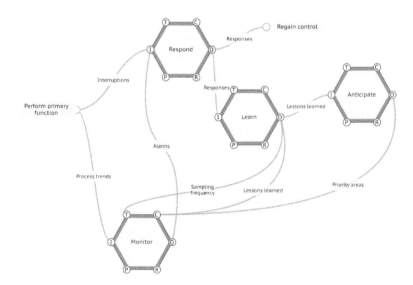

Figure 6.4 The basic model of the four potentials, using the FRAM.

Output (O), Preconditions (P), Resources (R), Control (C) and Time (T). In order to illustrate the systematism of the FRAM, the basic model has mainly described the Inputs and Outputs of the four functions that represent the four potentials. The basic model is, however, just the starting point for a more detailed, and therefore more realistic, model.

The detailed model

A basic FRAM model can be developed further by considering each of its functions in detail. This will very likely lead to the introduction of additional functions to be considered and so on. Even the first iteration of the model required two additional functions, as described above. (Lest the reader begins to worry about where this may lead, there is a built-in stop rule in the method.) The expansion of the basic model shown in Figure 6.4 will be illustrated by looking at the function <Learn>. In order to complete the model, the three other functions must, of course, also be expanded, but this is left as an exercise for the reader. (A possible solution is, however, provided by Figure 6.6.) The expansion is made by considering the six aspects of the function. This may in principle be done in any order.

- Output: The main Output from <Learn> is the [Lessons learned]. This is in the first iteration used as a generic term but may obviously be specified as different types of learning outcomes, for instance, with regard to what went well or what went badly, with regard to the ways in which the [Lessons learned] best can be used (design, instructions, training, communication, etc.). For the development of the model, the more interesting question is which other functions use or possibly depend on the Output of <Learn>. The [Lessons learned] may first be used to control <Monitor> and as a major Input to <Anticipate>. But the [Lessons learned] may also be the Input to two further functions. One, called <Update indicator list>, can use the [Lessons learned] to produce a set of [Key performance indicators] that in turn control <Monitor>. The other, called <Update and revise responses and monitoring>, can use the [Lessons learned] to generate the [Plans and procedures] that control <Respond> as well as to deliver the [Monitoring strategy] that controls <Monitor>. As already described, yet another important Output from <Learn> is [Sampling frequency], which specifies how often <Monitor> is carried out.
- Resources: In order to learn, it is necessary that some resources be available. In generic terms, these may be called [Staff and equipment]. (When a model is developed for a specific organisation, the resources can of course be described more precisely.) Resources in turn have to come from somewhere; they have to be the Output of one or more other functions. In this case, it is assumed that the Resources are produced by a function called <Manage

operational capabilities>. Failure to specify the origin of an aspect, in this case <Staff and equipment>, would result in an incomplete model.

- Control: The way in which an organisation learns must be systematic and planned, hence controlled. The Control is partly the [Learning strategy], which describes when learning should take place and how it should be done, and partly the [Business strategy], which sets the priorities for learning. The former can be seen as an Output from a function called <Ensure operational readiness>, while the latter may be an Output from a function called <Develop business strategy>. Both will initially be treated as background functions. (Another possibility would be to define [Business strategy] as an Output from <Anticipate>.)

- Time: Learning should ideally take place when the situation has reached a stable state, in order to avoid learning from something that may change later on. To be effective, <Learn> must be timed so that it does not happen before a stable situation has been established. This temporal condition relates to the [Latency of results], which can be proposed as an Output from a function that may be called <Reach equilibrium>. The latter is just a placeholder for the many things that are necessary for a temporary disturbance to settle and an equilibrium state to be achieved.

- Input: The main inputs to <Learn> are assumed to be the [Responses] to specific events and the [Process trends]. The [Responses] have already been introduced as an Output from<Respond> while [Process trends] can be seen as an Output from a generic function called <Perform primary functions>.

Many functions also have some Preconditions defined, i.e., conditions that must be fulfilled before the function can be carried out. In the current example of <Learn>, it is assumed that there are no special preconditions.

Altogether, this expansion of the basic model shows that the function <Learn> is coupled not only to the three other main functions but also to nine functions that as yet are undefined. The nine new functions describe both how the Output – or potentially Outputs – from <Learn> are used and where the various inputs and conditions that <Learn> needs come from. The detailed model of the potential to learn, or more precisely the first iteration of a more detailed model, is shown in Figure 6.5.

The complete model

The same procedure can be carried out for the other three potentials or functions <Respond>, <Monitor> and <Anticipate>. For each function, it should be carefully considered what the possible Inputs, Outputs, Preconditions, Resources, Controls and Time relations are. This may lead to the definition of a number of other functions required to make the model consistent. Some of these may already have been defined, while others are new. The whole procedure is explained in the Appendix. A possible result is shown in Figure 6.6. Although

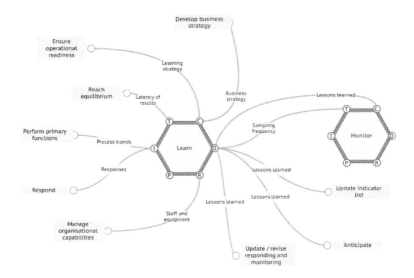

Figure 6.5 A detailed FRAM model of the function <Learn>.

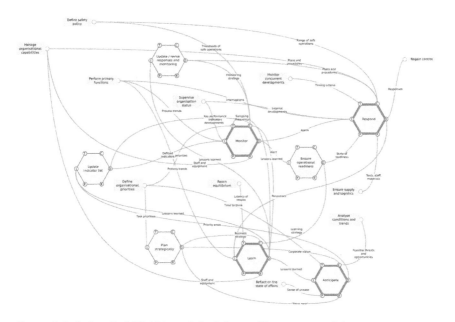

Figure 6.6 A detailed FRAM model of the resilience potentials.

this is not a final model of how the four potentials depend on each other, it does illustrate how it is possible to develop a generic model of such dependencies, and thereby provide an overall picture (literally) of how an organisation may be able to develop and maintain the resilience potentials. The reason why it is a generic rather than a final model is simply that it describes some 'common sense' dependencies rather than a specific organisation. The 'common sense' dependencies are all based on practical experience, and many of them may be found in practice. But the actual application of the RAG should be based on a specific rather than a generic model.

A generic model of resilience potentials

The model in Figure 6.6 shows how the four potentials functionally depend on each other and how the four potentials seen as functions require a number of other functions. But the four potentials are generic and do not represent the level where the practical assessment of an organisation's potentials for resilient performance can take place. They do not represent the level where the operational model or description of an organisation should be made or the level where the actual management is carried out.

In order to achieve that, it is necessary to refer to a specific organisation and a specific situation. That will make it possible to provide more concrete and detailed descriptions, in the same way as the four abilities in practice are assessed by the specific diagnostic questions. Just as the questions must be tailored to the organisation, so must the functional model.

The generic model is nevertheless helpful in thinking about specific initiatives or interventions to manage the potentials for resilient performance, both what they may require to work properly and what kind of secondary or derived consequences they may have. It is clear, even from the overall model, that it is inadvisable and probably inefficient to try to manage any of the four potentials or the many subfunctions that make up the four potentials one by one. Because the model provides a compact representation of the dependencies needed to fulfil the potentials, as well as how important or critical they are, it can also serve as a basis for defining and refining specific diagnostic questions.

Chapter 7

Developing resilience potentials

Changing an organisation's performance is never simple. The first challenge is to understand how an organisation functions 'internally'. In order to change an organisation's performance, and indeed in order simply to manage it, it is necessary to have a specific model of the organisation, a convenient description that represents the various organisational processes or 'mechanisms' that together produce the organisational performance. Most models are underspecified, both because they refer to generic rather than specific functions and because the internal dependencies rarely are explained or described in detail. Generic models are typically simple flow chart models where the nature of the flows is rarely specified in much detail and where the parts are organised as either a hierarchy or a network (cf., Figures 6.1 and 6.2). Specific models are usually organisational charts that show the relations between roles and departments, rather than how the organisation actually functions. Even for such models, people find it mentally or cognitively difficult to comprehend the potential consequences of changes since the reasoning usually is limited to simple first-order effects (trivial cause-effect relations) and conveniently disregards the indirect and therefore unanticipated consequences that arise from second-order effects and nontrivial interactions. Yet models where there is a simple linear dependency between the parts say very little about how an organisation functions and are by and large useless as a basis for planning and managing changes.

There is a second and more serious obstacle, namely, the question of what it more precisely is that determines how an organisation performs and how well it does relative to some criterion or criteria – such as safety, quality, productivity, customer satisfaction, ensuring availability, etc. This is, clearly, not unrelated to the first obstacle, especially if the model of the organisation goes beyond the mere structure and includes something about the 'causes' and the 'mechanisms'. The understanding of what determines performance is crucial because it is a necessary condition for management and control. Cybernetics formulated that in the 1950s as the Law of Requisite Variety (Ashby, 1956). The Law of Requisite Variety is concerned with the problem of regulation or control and expresses the principle that the variety of a controller should match the variety of the system to be controlled. Effective control is therefore not possible if the controller has less variety than the system. This has also been expressed by saying that 'every good regulator of a system must be a model of that system' (Conant and Ashby, 1970).

How (well) an organisation performs is of course inseparable from how (well) the people in the organisation perform. There are many ways to define what an organisation is, but common to them all is that an organisation is a collection of people that work together to pursue collective goals. (Indeed, if there are no people, but only the physical structures left – such as an empty factory – there is no organisational performance either.) The people are usually organised in a way that is assumed to contribute to the objective, the ongoing activity. The purpose of an organisation is to manage functions and responsibilities among units or groups, to allocate resources in the best possible way and to monitor and adjust performance over time. The performance of the organisation there-fore depends in a fundamental way on the performance of the people in the organisation.

This leads to two different positions on how to change an organisation's performance. One position is that what people do is determined by the organi-sation or by central aspects of the organisation, most notably the organisational culture or varieties thereof. The logical consequence is that people's perfor-mance can best be changed by changing the organisational culture and specifi-cally the safety culture – 'the way we do safety around here'. The solution is to focus on the organisation and try to change that. The other position is that the organisation's performance should be understood as a cumulative or aggregated (combined) effect of the performance of the individuals in the organisation. The consequence in this case is that the path goes through changing the individu-als' performance directly.

A third position is, of course, that the performance of an organisation and the performance of individuals are inextricably coupled so that it does not make sense to refer to one without referring to the other. This is the position taken by resilience engineering.

Changing organisational culture

Focusing on organisational culture as the critical issue can be seen as yet another case of monolithic thinking and reasoning. It is certainly convenient to explain organisational performance in terms of a single factor or cause and to focus on changing just that. If culture is the predominant cause, then it must be more efficient (and easier too) to use the culture to change the practice than doing it the other way around. This is the rationale behind the idea of a safety journey and the scheme to achieve safety by working on the attitudes of people – by winning their 'hearts and minds' (H&M) (Parker, Lawrie and Hudson, 2006). The reputed success of this approach is unfortunately based on a false ana-logy. The 'H&M' was taken from a campaign to stop smoking (Prochaska and DiClemente, 1983). Smoking is fundamentally what an individual does. There is very little in the nature of regular or regulated performance, hence little possibility of changing that. (The 'H&M' may be seen as corresponding to the notion of espoused values. In that sense, there may be some reasonableness to it, but that is not what the proponents argue. Indeed, they fail to see it as a composite or aggregate issue but stick to the convenience of monolithic thinking.)

The 'H&M' analogy is problematic because it is too simple, at least in the way it is commonly used. To smoke or not to smoke is an uncomplicated, unitary activity. The decision to smoke or not to smoke is assumed to be uncomplicated, to wit a decision about doing something or not doing it – even allowing for the existence of social pressures. This decision is in turn based on personal preferences or attitudes. It therefore makes sense that if these attitudes are changed, then the person will no longer smoke.

This analogy is misleading because everyday work cannot be reduced to a question of whether to do something or not to do something. Everyday work is rarely about making binary choices. In fact, it is more about creating the opportunities or alternatives for doing something, to make sense of a situation in order to know what to do – as in naturalistic or recognition-primed decision making. It is a question of *how* to do something, not *whether* to do it. Figuring out how to do something is furthermore part of a context or a continuum, rather than an independent action (such as lighting a cigarette). Smoking is rarely part of work but is rather an interruption of work. But actions, steps in everyday work, are by their very nature part of something and not isolated; we do not think about them as individual steps.

This is why the H&M analogy fails. By the same argument, any other approach or scheme or solution that relies on monolithic reasoning and assumes a single dominating factor will also fail. The actions that together constitute our work are not determined only – or even mostly – by attitudes and beliefs. They are determined by the actual and practical necessities of work. Therefore, if we want to change what people do, we should not rely on changing their espoused values (H&M) as the primary solution. We can change what people do by changing the determinants of work, the reasons why they do things. This might be the balance of demands and resources, the design of the workplace and the interfaces, the way that activities are supported – or hindered, the expectations that are given in the procedures and guidelines as well as the social expectations by peers and of course also the attitudes, but never the attitudes alone. Indeed, one might hypothesise that the attitudes would change when the practice changes, rather than the other way around.

Changing practice

The alternative to changing the culture is to change the practice. One approach is to focus on the individual and on what it is that makes people do what they do. This has already been discussed in Chapter 3, where a number of theories about individual behaviour (in an organisation) were presented.

Since people reflect on what they do, a change in practice will also gradually lead to a change in attitudes, in espoused values and assumptions. The advantage is that it is much easier to change practices than to change attitudes and values and much easier to notice whether it works – the effects are direct rather than indirect. In building the resilience potentials, the change is to the ways in which people – and through them the organisation – respond, monitor, learn and anticipate.

The performance of an individual is, however, not independent of the situation, the operating conditions and the social environment. The basis for changing a person's performance must acknowledge the person's understanding of his/her own performance, of the performance of the people around him/her (the work group or close collaborators) and of the performance of the organisation. In any given situation, a person will do what he/she thinks is the right thing to do in order to ensure that the intended consequences are obtained. One of these consequences is of course the factual outcome or result of the activity. But there are others such as the acceptance or approval of others (colleagues, peers and perhaps managers). Usually people will try to avoid a conflict with the general expectations of performance – unless, of course, they intend to do harm to others or instil terror and fear. The 'others' may grow to go beyond the social group and include the higher echelons of the organisation or perhaps even represent the 'organisation' itself. In the latter case it will, of course, be the person's imagination of what the response would be rather than an actual response. And a person's imagination or expectations to what the response will be is in a sense the espoused values, since these represent the accepted norms of 'how we do things here' – as the popular definition of safety culture goes. The understanding of the performance of the organisation precisely refers to the common values that characterise the organisation (the shared basic assumptions), hence perhaps to something that is as close to the notion of an organisational culture as one can reasonably come. Indeed, while the understanding of one's own performance as individual performance refers to the person's own goals and expectations (criteria or *anspruchsniveau*), the understanding of one's own performance in the context of the organisation, of the greater whole, the greater common good, must refer to some kind of shared values and shared norms. The social norms are not only a reference for what each person does in his/her own right but also the basis for what people expect others to do.

The third way

It is tempting to think in terms of simple alternatives, one being whether individual performance can be changed by means of the organisational culture (or safety culture or even a culture of resilience) and the other being whether organisational culture can be changed by working on individual performance and behaviour. The juxtaposition is, however, too simple and therefore too good to be true. As Chapter 3 concluded, it is not enough to change the organisational culture in order to change what people do. Indeed, neither safety culture nor any other single factor will suffice. Monolithic thinking is cognitively appealing but also grossly inaccurate. Organisational culture is not a single or unitary concept but entails a distinction among – at least – three parts: the artefacts, the espoused values and the shared basic assumptions (Schein, 1990). The organisational culture should be thought of as the intersection of these rather than as a factor in its own right, cf., Figure 7.1.

Figure 7.1 illustrates how organisational culture can be represented as the intersection of artefacts, basic assumptions and espoused values, hence that it is a composite rather than a simple or monolithic concept. But Figure 7.1 is not a model of organisational culture because it does not make clear how the constituent parts – artefacts, espoused values and basic assumptions – work together. It can therefore not be used directly as a basis for planning and managing organisational change.

Chapter 5 argued that the four potentials could be used as a starting point for a generic model of how an organisation can perform in a resilient manner. This, of course, is a model of how an organisation can sustain its performance in general and not a model of what the organisation does or of what it provides – e.g., a hospital, a coffee shop, a passenger ferry, a supermarket or an oil rig. It must therefore be supplemented by, or combined with, a model of how an organisation manages its primary activities. While this could be done by one of the common model types shown in Chapter 6, it would make good sense to use a functional model instead. The FRAM is well suited for that.

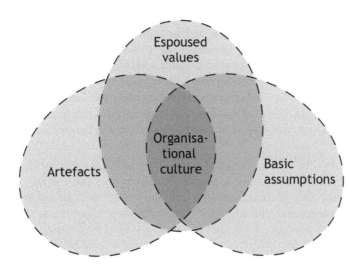

Figure 7.1 Organisational culture as the intersection of artefacts, espoused values and basic assumptions.

'Dysfunctional' and 'resilient' organisations

As a starting point for considering how an organisation can build its resilience potentials, consider two extreme cases: an organisation that functions rather badly (a 'dysfunctional' organisation) and an organisation that functions rather well (a 'resilient' organisation).

- The first case is represented by an organisation that goes about its business in a routine manner, whose only 'claim' to resilient performance is that it is able to respond when something (unexpected) happens. While such an organisation is imaginable (such as a financial institution that is 'too large to fail'), it cannot survive for long unless it exists in a nearly stable operating environment where routine responses suffice. When a dysfunctional organisation responds, it does so in a stereotypical manner. It does not have the potentials to monitor, learn or anticipate. The lack of (effective) monitoring means that it is never prepared when something happens and that everything therefore is a surprise. Because the operating environment is stable, an organisation may little by little get used to the surprises and in a primitive sense learn which responses are required. However, an organisation that does not learn properly is limited to the initial set of responses even if it becomes more adept at executing them – almost as in a conditional reflex. The potential to respond is fundamental since an organisation (a system, an organism) that is unable to do so with reasonable effectiveness sooner or later will become extinct or 'die' – in some cases literally.
- The second case is an organisation that has the potentials to respond, monitor, learn and anticipate. A 'resilient' organisation will respond in an efficient and flexible manner; it will be aware of what happens inside it as well as in the operating environment; it can learn effectively from past experiences and it can consider possible conditions or 'futures' beyond the current situation. It is furthermore able to do all of this acceptably well and to manage the required efforts and resources appropriately. It is never satisfied, since it realises that the future is uncertain and that it is an advantage to have a permanent sense of uncertainty – never to become complacent.

Given the two extremes, a crucial question is how an organisation can improve its resilience potentials in practice so that it does not end by becoming dysfunctional. One position could be that since there are four potentials that need to be considered, the answer is to develop them all as much as possible. But this raises an issue of whether it should be done by working at them all at the same time, in parallel, or whether it should be done by improving them one by one and then whether there is preferred order or priority. In practice, it will be impossible to consider them all at the same time because this will require an inordinate amount of resources (human and financial), because the four potentials are very different and 'grow' or develop at very different 'rates' and because their relative importance depends on the characteristics of an organisation. Neither does it seem reasonable to address them randomly because they are mutually dependent or coupled (cf., Chapter 6 and Figure 6.4). In this case, the mutual dependencies are, however, not a limitation but an advantage because they can be used to propose a strategy for developing an organisation's resilience potentials that is more sensible – and therefore also more effective – than any other.

In the worst case, the development of the resilience potentials has to start from a dysfunctional organisation that has the potential to respond but none of the three others. It is in fact possible to imagine an even more extreme case, namely, an inert organisation that is impervious to what happens around it. While both types of organisation are theoretically possible, neither is likely to be found in practice. In order to exist and survive, an organisation must have at least a rudimentary potential to monitor. A further type is an organisation that keeps behaving in the same manner, despite the availability of feasible alternatives and contrary to own interests. Incredible as it may sound, such organisations have existed throughout history and even exist today (Tuchman, 1985).

Developing the potential to monitor

The starting point for developing the resilience potentials cannot be an organisation that is unable to respond, since such an organisation would need more drastic measures. The starting point must be an organisation that can respond and therefore is able to survive. It is, of course, always possible to improve the potential to respond. Even in cases where the operating environment is stable and completely predictable, the potential to respond can be improved with regard to the speed of responding or by fine-tuning the triggering conditions, for instance. But when the operating environment, as is usually the case, is incompletely predictable then responding depends critically on the potential to monitor. Rather than focusing narrowly on strengthening and improving the potential to respond, a better strategy is to develop the potential to monitor. Monitoring allows an organisation to keep track of how the operating environment changes and to detect developments and disturbances before they become so large that a response becomes necessary. This will, on the one hand, enable the organisation to prepare itself for a response, for instance, by reallocating internal resources or by changing its mode of operation, to shut down some services and to active others, and on the other hand, allow responses to be made to even 'weak signals' – before a situation has deteriorated. Responding at an early stage in the development of an event will generally require fewer resources and take less time, although it also incurs the risk that the response may be inappropriate or even unnecessary. The trick is obviously to avoid false positives, where a response is started even though the situation does not need it, as well as false negatives, where a response is not started even though it is needed. Taking these risks is preferable to operating in a reactive mode only.

Reactive and proactive adjustments

The key feature of a resilient organisation is the ability to adjust how it functions. Adjustments can in principle take place either after something has happened (be reactive, responding to feedback) or take place before something happens (be anticipatory or proactive, controlled by feedforward) based on calculations or assumptions about what will happen in the future – either in the short run or the long run.

Reactive adjustments are by far the most common. For instance, if there is a major accident in a community, such as a large fire or an explosion, local responders will change their state of functioning and prepare for the many different types of consequences that may follow. Responding when something has happened is, however, insufficient to guarantee an organisation's safety and survivability. One reason is that an organisation can only be prepared to respond to a limited set of events or conditions and usually only for a limited duration. Another reason is that if a response waits until the need is obvious, the damage will have had time to grow and spread.

Proactive adjustment means that the system can change from a state of normal operation to a state of heightened readiness before something happens. In a state of readiness, resources are allocated to match the needs of the expected event, and special functions may be activated. A trivial example from the world of aviation is to secure seat belts before start and landing or during turbulence or the preparations made before a tornado or a hurricane – a violent storm – hits an area. In this case, the future events are consequences of regular, perhaps even scheduled, activities, hence highly predictable. In other cases, the criteria for changing from a normal state to a state of readiness may be less obvious either because of a lack of experience, because the future is uncertain, because the validity of indicators is questionable or because the signals are 'weak'.

Alternative route: learning before monitoring

While developing the potential to monitor effectively enhances the potential to respond, it is also necessary briefly to consider two other possibilities as a 'first' step, namely, developing the potential to learn and developing the potential to anticipate. Learning is, of course, important for the potential to respond, since it is through learning that the responses become matched to the characteristics of the operating environment. While learning may make the responses more versatile, an inability to monitor will still limit an organisation to being reactive, hence at risk of responding too late and gradually lagging behind what happens around it. Prioritising learning over monitoring implies a belief – or faith – that what happens in the future mainly will be a kind of repetition of what has happened in the past. This is typically the case for organisations that only learn from accidents by adhering to a 'find and fix' strategy. Fixing something once a cause has been determined will obviously only have the intended effect if the operating environment remains the same. Ineffective monitoring must be compensated by maintaining a high state of readiness, which means that resources and capabilities constantly are reserved for events that may possibly happen even if the situation is completely different. Simply improving the potential to respond by prioritising learning over monitoring incurs a cost, which makes this approach self-defeating in the long run.

Alternative route: anticipation before monitoring

Prioritising the potential to anticipate over the potential to monitor is not a good choice either. The anticipation of future developments and changes can indeed be the basis for proposing new types of responses (and capabilities), but if monitoring is ineffective the mode of responding will still be reactive. As in the case of prioritising learning over monitoring, prioritising anticipating over monitoring is only justified if external events are very infrequent. One example could be a company with a stable production, either a coal mine with a well-functioning operation in a large and homogeneous coalfield or a production line for a product that historically is selling well. In such cases, it is possible that the existing sets of responses – and the monitoring – are adequate and that production, whatever it is, can go on in a steady fashion. Under these conditions, it might be more advantageous to improve the potential to *anticipate* whether there may be changes in the market, whether new customer needs develop and whether new regulatory developments are likely, rather than to monitor.

Developing the potential to learn

For an organisation that already is able to respond and to monitor, the next logical step is to focus on the potential to learn – while still paying attention to the potentials to respond and monitor. Learning is necessary for several reasons. The most obvious is that the operating environment always is in a flux, which means that there always will be new and unexpected situations or conditions. It is important to learn from these and in particular to look for regularities that can improve the potentials to respond and to monitor. Another important reason is that the potential to respond always will be limited. It is neither possible nor affordable to prepare a response for every event or for every possible set of conditions. This means that an organisation every now and then will find itself in a situation where it does not know how to respond. It is clearly important to learn from these situations, to evaluate whether they are unique or likely to occur again and to use that to improve both responding and monitoring. But it is just as important, if not more important, to learn from responses that went well. An organisation can use this experience to improve the precision of responses, the response time, the set of cues or indicators that are monitored, etc.

Simply accepting things that go well is tantamount to being complacent and thereby missing an essential opportunity to make improvements.

Alternative route: anticipation before learning

While it makes good sense at this stage to develop the potential to learn rather than the potential to anticipate, it is still possible to consider the alternative, namely, to develop the potential to anticipate rather than to learn. One argument against doing things in this order is that effective anticipation depends on learning. Anticipation is the 'disciplined imagination' that is used to consider possible future worlds. Anticipation should consider possible changes to market conditions, new regulatory demands, new technologies, political upheavals, natural and environmental disasters, pandemics, etc. Yet without the potential to learn, the 'imagination' runs the risk of becoming undisciplined or at least uncalibrated. Since the use of anticipation to manage the organisation – be it in the business or the safety directions – entails taking a risk, anticipation without learning is probably not a good idea.

Developing the potential to anticipate

At this point, when an organisation can respond, monitor and learn, all that remains is to develop the potential to anticipate. The value of anticipation has already been discussed. Anticipation is not just an extrapolation from the current situation. Anticipation can more precisely be used to enhance the potentials to monitor (suggesting which indicators and priority areas to look for), to respond (outlining possible future scenarios) and to learn (prioritising different lessons). Learning can be used to improve the potential to respond, to select appropriate indicators and cues and also to hone the imagination that provides the basis of anticipation. Monitoring can primarily be used to improve the potential to respond (increased readiness, preventive responses). Responding can also provide the experience that is necessary to improve learning as well as anticipation.

Choosing how to develop the resilience potentials

This altogether means that an organisation that wants to improve its resilience potentials must carefully choose how and when to develop the four potentials – in particular whether to concentrate on one or to work on two or more at the same time. This is where the use of the Resilience Assessment Grid (RAG) can play a role. An organisation must first determine how well it does with regard to each of the four potentials – by carefully assessing the functions that contribute to each potential – and then plan how to go about developing them. In doing so, it is necessary to take the mutual dependencies into account, as well as what means are best suited, cf., Chapter 6. In some cases, a potential or its constituent functions may be developed via technical improvements such as better sensors or more powerful ways of analysing measurements. In others, human factors or organisational relations may be more important. There may finally be cases where attitudes, or even something akin to safety culture, are of instrumental value.

Developing resilience potentials means establishing (and nurturing) a set of attitudes that is favourable to Safety-II – that emphasise the importance of making performance better rather making it less worse. In other words, attitudes that encourage people to think and focus on what goes right rather than just what goes wrong, that make them notice (perceive) the things that otherwise cannot be seen, that make them strive to do things better rather than to prevent doing things wrongly, etc.

In terms of setting the goal, resilience engineering does not prescribe a final solution. Instead, each organisation needs to decide how far it must develop its resilience potentials, expressed as differential levels of the four potentials. This is a pragmatic rather than a normative choice and depends heavily on what the organisation does and in which context. (It obviously also depends on an organisation's potential to anticipate and on the level of ambition for being prepared for the future.) Unlike the notion of safety culture, as it is expressed in the five levels model, there is no ceiling for the resilience potentials. Responding, monitoring, learning and anticipating can always be improved just as an organisation always can become better at what it does – in terms of productivity, safety, quality, customer satisfaction, etc.

Managing the resilience potentials

In the tradition of monolithic thinking, safety is usually referred to as a single concept or quality. This has resulted in the custom of referring to a 'safety culture', to 'safety management' and to 'safety management systems' and assuming that changing these will bring about the desired outcome.

Safety management systems typically emphasise a single issue, not only in terms of indications or manifestations (failures or the lack of safety) but also in terms of approach. From a Safety-I perspective, the purpose of safety management is the reduction and possible elimination of adverse outcomes, and the approach must therefore be based on an understanding of how adverse outcomes happen. The second principle of the *causality credo* implies a value congruence between causes and effects – in the sense that an adverse effect must be due to an adverse cause. The understanding of how failures and malfunctions happen therefore becomes critical. The preferred explanations have historically gone from the technology, to the human factor and then to the organisation and safety management, as described in many places.

It is tempting and sometimes irresistible to suggest that the way to increase safety goes through improved resilience and therefore to propose various ways of supporting resilience either technological or organisational. Such an endeavour is, however, going to fail for the simple reason that resilience is neither homogeneous nor a single concept, a single quality or a single capability. As argued in Chapter 2, there is no single or simple quality called resilience, and it makes little practical sense to ask whether an organisation can have resilience or even be resilient. There is therefore no way in which the resilience of an organisation as such can be improved or supported. It is more useful to propose that an organisation can have the potentials for resilient performance, and these potentials can be managed – and indeed should be managed. The main inconvenience is that we regrettably end up with the rather unwieldy term 'management of the potentials for resilient performance' or the shorter 'management of resilience potentials' instead of having a nice term such as 'resilience management'.

Instead, we must consider how it is possible to support or facilitate the four potentials as placeholders for operational or concrete aspects of performance. This need not address all four at the same time, as described in this chapter. It is quite reasonable to look for ways to improve an organisation's potential to monitor, for instance. But in doing so, it is crucial to keep in mind that the four potentials are coupled in a nontrivial manner. If the focus is on the ability to monitor, this goal should not be pursued without keeping in mind what the potential to monitor depends on and requires and how a change here may affect the three other potentials. The general or generic dependencies have been outlined in Chapter 6, but in any concrete case they must of course be described more precisely.

Using the RAG

Resilience engineering does not prescribe a certain balance or proportion among the four potentials. The proper balance must be based on knowledge of what an organisation is supposed to do and experience from how well it is able to do that. Since this must depend on the domain, it is impossible to propose a 'standard' value. For a fire brigade, for instance, it is more important to be able to respond than to anticipate. For a sales organisation, the potential to anticipate may be just as important as the potential to respond. But resilience engineering does make clear that it is necessary for an organisation to possess each of these potentials to some extent, in order to have the potential for resilient performance. All organisations traditionally put some effort into the potential to respond. Many also put some effort into the potential to learn, although it often is in a very stereotyped manner. Fewer organisations make a sustained effort to monitor, particularly if there has been a long period of stability. And very few organisations put any serious effort into the potential to anticipate.

To summarise, there are four important points to remember when using the RAG, as described in Chapter 5.

- Develop four sets of diagnostic and formative questions that are specific to the organisation.
- Develop a FRAM model of the functional couplings of the four potentials for the specific organisation. If possible, also develop a FRAM model of the organisation's primary functions.

- Establish a core group of respondents that have practical experience with how the organisation works and that will be available for repeated assessments.
- Use the RAG for repeated assessments with intervals that are appropriate for the organisation and its primary functions. Use the assessments to manage and improve the organisation's resilience potentials.

The diagnostic questions should comprise issues that refer to how people perform as well as to performance characteristics of the organisation. The use of the RAG therefore illustrates the third way.

Chapter 8

The changing face of safety

It is easy to understand why safety historically has been associated with the absence of or freedom from harm and injury. When we suffer harm, whether as individuals, groups or societies, it is noticed both because it is unexpected and unusual and because it results in discomfort, pain or even loss of life or property. All living organisms respond to harmful stimuli, and in the case of humans the response is not only physiological but also psychological and societal. We know that when we have suffered harm or injury, we try to find and understand the cause(s), and we try to categorise the conditions so that we can remember them and learn to recognise them. We do this as individuals, groups and societies. Since the survival value of that is indisputable, it is little wonder that safety efforts have been focused on the elimination or prevention of hazards and risks. Neither is it strange that safety efforts have been, or even must be, to a large extent reactive. When something goes wrong, the natural instinct is to respond immediately, usually by getting out of harm's way. Most safety paradigms therefore start with a harmful event. That is followed by an assessment of the level of harm, by an attempt to identify the causes, by a search for solutions and finally by implementing the solutions after having ascertained (as far as possible) that they will bring about the desired effects.

In the case of regularly occurring events, the eagerly sought 'freedom from unacceptable risk' can usually be achieved in five different ways either alone or in combination (cf. Table 8.1). The first and most obvious way is by elimination, by removing the offending activity or 'component' from the way the system works. A second way is by redesign, where the focus can be either the humans, their competence and the way they work; the technology or the organisation. A third way is by prevention, by introducing active or passive barriers that prevent the initiating event from occurring. A fourth way is by improving monitoring in order to reduce the level of surprise when something happens – and preferably to eliminate the element of surprise completely. A fifth and final way is by protecting against the consequences when something happens. The modern car can serve as a good illustration of all five. Drivers are gradually being eliminated or taken out of the loop (replaced by technology) because they are seen as failure prone. Redesign is also widely used, both of the driver's workplace, of individual cars and of the traffic system. Prevention uses a variety of passive and increasingly active barriers such as crumble zones, brake

Table 8.1 Possible reactions to 'unacceptable risks'

Response	Detailed response	Examples
Eliminate	Remove	
Redesign	Human	Training, replace by automation, task allocation
	Technology	Improved design, improved components, displays and controls
	Organisation	Safety culture, communication, allocation
Prevent	Physical barrier	Walls, fences, railings, bars, cages, gates
	Functional barrier	Lock, interlock, password, etc.
	Symbolic barrier	Warnings, warnings devices, interface layout, signs, symbols
	Incorporeal barrier	Rules, guidelines, safety principles, restrictions and laws
Monitor	Online (synchronous)	Measurements, key performance indicators
	Offline (asynchronous)	Test, inspection, event reporting
Protect	Automatic	Fail-safe devices
	Managed	Emergency response, 'firefighting'

assistance, traction control, monitoring of traffic conditions and of driver states, etc. And drivers are finally protected by safety belts, air bags, safety cages, etc.

There is, of course, also the possibility of not doing anything; this sometimes happens if the risk – or rather, the loss – is considered acceptable, or if it is prohibitively expensive to do something.

In the case of irregular events there is little that can be done. The reason is mainly economical. While it is cost-effective to be prepared to respond for regular events and to try to prevent them from taking place, irregular events simply do not occur often enough to justify the necessary investment. The situation is even worse for unexampled events such as Chernobyl or Fukushima. Here extensive post-mortem analyses may provide some psychological comfort but little in the way of substantial improvements.

While the response to accidents usually is presented as a rational, engineering concern, it cannot hide the fact that there is an emotional or affective concern as well. When something bad happens, there is a need to *feel* safe, which sometimes may dominate the need to *be* safe. This is seen not least in the 'official' response to accidents, such as "the government says it will leave no stone unturned in its quest to find out exactly what caused the deadly bus accident". This basically means that 'we' will do everything 'we' can so that 'you' can feel safe.

An example is provided by the problems with the overheating of the lithium-ion batteries on the Boeing 787 Dreamliner. During December 2012 and January 2013, a number of aircraft had problems with batteries, which were either damaged or caught fire. On January 16, the Federal Aviation Administration (FAA) issued an emergency airworthiness directive grounding U.S.-based Boeing 787s. Boeing immediately tried to find the cause of this. In March 2013, after more than 500 engineers together with outside experts had spent more than 200,000 h of analysis, engineering work and tests, the Dreamliner's chief engineer acknowledged that Boeing had not found the exact cause of the overheating and that it probably never would. They had examined 80 potential problems that could lead to a battery fire, grouped them into four categories and designed solutions for each category. In April 2013, the FAA approved Boeing's modifications and later approved the aircraft for flight. According to the Transportation Secretary, 'Safety of the traveling public is our number one priority. These changes to the 787 battery will ensure the safety of the aircraft and its passengers'.

The traditional meaning of safety (Safety-I) is thus 'without' – meaning that we are safe if we are without the adverse events (failures or malfunctions) or without the negative consequences (harm or injury). In consequence of that, the main efforts have been to ensure that a condition of 'without' could be established and maintained by elimination, prevention and protection as summarised above. The epitome of that is, of course, the principle of 'defence in-depth', where barrier is put upon barrier to stop the penetration of harmful influences. (The origin is in the use of physical defences such as walls and moats to protect cities or castles – or their modern day equivalent of the Maginot line built by France in the 1930s to slow down the progress of an enemy (Germany) and to give their army time to mobilise in the event of attack. But the term 'defence-in-depth' has been extended to describe the use of barriers in general, as elegantly illustrated by the Swiss cheese model.) This approach has worked acceptably well for centuries, as long as sociotechnical environments, for lack of a better term, were stable and predictable. Put differently, the rate of change and the rate of invention were for a long time so slow that it was (easily) manageable. But after the middle of the 20th century, this was no longer the case. The change was due to a combination of two factors, one being human inventiveness and the other being the constant striving to increase our mastery of the world around us. It is ironic that the combined effect of the two factors leads to a dynamically unstable situation where we introduce solutions that we are unable to control in order to control systems that we do not understand in the first place. In other words, we use the power of technology to compensate for our inability to master what we build. The dilemma in the design of sociotechnical systems is that it attempts to solve the problems of today with the mindset – the models, theories and methods – of yesterday that inadvertently creates the complexity of tomorrow.

Many people in many different fields of activity have gradually realised that this situation is untenable and that it therefore is necessary to look for a different solution. In hindsight, this realisation was – partly, but not completely – the motivation for the development of resilience engineering, although the dilemma (obviously) was not formulated as clearly at the time.

Safety as a privative

In grammar, a privative is a particle that negates or inverts the value of the stem of the word. For example, **un-** in unprecedented or **in-** in incapable. Using this as an analogy, Safety-I can be seen as a privative because it is defined by its opposite, by a state where it is absent or by a lack of safety. This has also been recognised by Reason (2000) as the safety paradox. But being safe, being free from harm and injury, means the absence of a 'lack of safety' rather than of 'safety'. This creates another paradox or absurdity, namely, that we try to learn more about safety by studying situations where we acknowledge that there was no safety – by 'learning from accidents'. Safety science is fortunately alone in taking this peculiar approach (Hollnagel, 2014b). Sciences in general try to study their chosen phenomenon in situations where it is present.

A physical analogy of a privative is *cold*. Any physicist or engineer can tell you that there is no such thing as cold; there is only lack of heat. Even though we may say that we shut the door to keep the cold out, there is no cold to keep out. Instead, we shut the door to keep the heat in. If we feel cold, we cannot become warmer by reducing the cold, i.e., by having less cold. We can only become warmer (stop freezing) by increasing the heat. For safety, there is an analogous situation. We cannot increase safety by reducing the number of accidents because the accidents represent a lack of safety. We can only get rid of accidents by doing things right more often. Measuring cold corresponds to measuring safety by counting accidents. Measuring heat corresponds to measuring what goes well.

One way of overcoming this problem is to offer Safety-II as that which we are without when we are unsafe, when there is an accident or an incident. A simple and elegant way of expressing this is to paraphrase WHO's definition of health as 'a state of complete physical, mental, and social well-being and not merely the absence of disease or infirmity'. Applied to safety, the paraphrase would be that safety is 'the ability of an organisation to perform as required under both expected and unexpected conditions and not merely the absence of unwanted outcomes'.

In the physical analogy, Safety-II corresponds to the heat while Safety-I corresponds to the cold. The juxtaposition of Safety-I and Safety-II is useful because it makes clear that there are two different phenomena or concepts – one corresponding to 'without' and the other corresponding to 'with'. While the solution works, it also introduces a problem, namely, that 'safety' in Safety-I and Safety-II is both a homograph and a homophone; yet it is not a homonym (cf., Figure 8.1).

Figure 8.1 The ambiguous nature of safety.

This situation is clearly neither desirable nor practical. It can also easily lead to confusion in discussions, since it may be difficult to know whether someone who uses the term 'safety' refers to the Safety-I interpretation or the Safety-II interpretation. The easiest way out of this problem would be to find a different word to represent what Safety-II stands for. The use of the terms Safety-I and Safety-II, however, has considerable rhetorical value because it makes clear that the traditional understanding of safety is not the only one. Yet in the long term, it becomes cumbersome to use a term and then immediately add that it is actually used with a completely different meaning.

(I am grateful to Professor Harold Thimbleby for pointing out that safety can be seen as a privative.)

Synesis

There is, fortunately, a term that means 'with' – not as in 'with harm and injury' but as in 'with positive or desired outcomes'. This is the word *synesis*, a traditional grammatical/rhetorical term derived from Greek σύνεσις (originally meaning 'unification, meeting, sense, conscience, insight, realization, mind, reason'). Although *synesis* primarily is used to describe a grammatical construction, the term makes good sense also in relation to Safety-II. Here *synesis* can be defined as the condition where multiple activities work together to produce or deliver acceptable outcomes. The individual activities may be partly incongruous or conflicting, but any difficulties are overcome by *synesis*, by the synthesis of activities that is necessary in order that today's sociotechnical systems can function as intended and desired. Thus, if we speak of the *synesis* of a clinic, or of a construction site, we mean the mutually dependent set of functions that are necessary for the workplace to carry out its activities as intended with respect to a set of relevant criteria such as efficiency (with a reasonable use of resources), reliability (it is highly predictable) and with acceptable quality. Note that there is no need to mention safety explicitly; the very fact that performance goes well means that there are no adverse outcomes. By using the term *synesis*, safety is thus defined as 'with' rather than as 'without'.

Synesis can also help to overcome a semantic problem. Many domains, health care being a prominent example but by no means the only one, conflate the terms safety and quality. In some cases, quality is seen as a component of safety, while in other cases, the opposite relation is assumed to hold. The same goes for safety and productivity, productivity and quality and so on. We can look at a process or work situation from a safety point of view, from a quality point of view, from a productivity point of view, etc. But in each case, we should keep in mind that any specific perspective only reveals part of what is going on, and that it is important to understand all of what goes on.

Since synesis clearly is a better term than Safety-II, it will be used throughout the remaining pages of this book.

The changing face of measurements

The changing face of safety also means that we need to find an appropriate way of measuring safety – or rather of measuring synesis. From Lord Kelvin's famous statement that you can only know something about that which you can measure to the cybernetic Law of Requisite Variety, it is clear that in order to manage something it is necessary to know what it is – to have a model of it – and to be able somehow to measure it. In Safety-I, the measures have traditionally been in terms of the number of unwanted outcomes (accidents, incidents, Lost Time Injury, etc.), thus leading to the ironic situation that a *higher* level of Safety-I corresponds to a *lower* value of whatever measure is used.

The conventional way of measuring safety can be expressed like this:

$$Safety = \sum_{1}^{n} \left(Adverse\ outcomes_i \right)$$

The adverse outcomes that are counted are naturally enough assumed to be the result or effect of a known cause or causes; of some kind of failures, malfunctions, shortcomings or of unknown or uncontrolled risks and hazards. The advantage of this way of measuring safety is that adverse outcomes are easy to detect and therefore also to count. The disadvantage from a management and control engineering point of view is that as time goes by and Safety-I improves, there will be less and less to measure, hence less and less feedback or information/data to support effective management. Indeed, the combination of 'learning from failure' and the 'zero harm' or 'zero accident' vision leads to a situation where there is nothing to learn from, hence no basis for improvement.

In 2001, the American organisational theorist Karl Weick famously proposed that safety could be understood as a 'dynamic nonevent', i.e., as the (adverse) outcomes that did not happen or were avoided rather than as the (adverse) outcomes that happened (Weick, 1987). The reformulation makes immediate sense, since the goal of safety management should be to ensure that there is an absence of adverse outcomes – or in other words that things go well. This logically leads to the following as a way of measuring safety:

$$Safety = \sum_1^n \left(-Adverse\ outcomes_i\right)$$

This definition means that safety is characterised by the absence rather than the presence of failures (things that go wrong) but that this absence is the result of active and continuing efforts. The practical problem is that safety in this way is expressed in terms of something that can neither be observed nor be measured. We can count the number of people who die in traffic accidents, for instance, or the number of times a train does not stop at a signal (SPAD or Signals Passed at Danger) in a railway network. But it is impossible to count the nonevents. In the case of traffic deaths, there is no meaningful way to count the number of people who could have died in traffic accidents but did not. In the case of SPAD, it may in principle be possible to count the number of times a train stopped at a signal – for a given region and a given period – but it is highly impractical, not least because there currently is no interest in knowing what this number is.

The definition of safety as the number of dynamic nonevents is attractive and in principle (or in spirit) in good agreement with resilience engineering, which from the start has emphasised that 'failure is the flip side of success'. But it would be even better, and in complete agreement with the definition of Safety-II, to think of safety in terms of 'dynamic events', meaning the events or activities that go well. And it is straightforward to propose a corresponding measure of safety, namely, as the number of acceptable outcomes. This would also mean that measurements of safety increase as safety gets higher.

$$Safety = \sum_1^n \left(Acceptable\ outcomes_i\right)$$

Safety, i.e., synesis, is the presence of acceptable outcomes. The more there are, the safer the system is. All that is required is to propose meaningful categories of successful outcomes. Although this at first may seem to be difficult because it is something that is unfamiliar to traditional safety management, it is not really so. Indeed, when other aspects of an organisation are managed – such as productivity, quality, customer satisfaction, etc. – it is quite natural to do so.

Product and process measures

While measures of recognisable outcomes, acceptable and unacceptable alike, are convenient and usually easy to get, they hide a major conceptual problem. Namely, what are the 'mechanisms' that bring about the outcome? In other words, what are the explanations (theories, models) that can be used to account for the outcomes that are measured?

The problem is illustrated by Figure 8.2. We assume, as indeed we must, that the outcomes in some way are the consequence of how an organisation functions, and we also assume that we can affect how an organisation functions by controlling it in some way. The assumptions can be relatively simple, for instance, the belief in linear causality that is part of all Safety-I models and of most organisational models as well, including the strategy map in balanced scorecards. The assumptions can also be more elaborate, such as in high reliability organisations or even nonlinear as in complex adaptive systems or resilience engineering.

Measures that are based on outcomes or products are usually simple to make, but their meaningfulness depends on the underlying model or assumptions about how an organisation functions. Since the relationship between the organisation (the process) and the outcomes (the products) often is underspecified or expressed in very general terms, as in the case of safety culture, for instance, outcome measures are usually not the best basis for managing an organisation. There are also other problems with outcome measures, not least that there may be a considerable lag or delay before they occur.

The example of the hospital standardised mortality ratio (HSMR) (Chapter 5) illustrated some of the problems with outcome measures. A more serious issue is that the relation between functioning and outcomes rarely is trivial. Thus, an 'unsafe' organisation may not have any accidents even for a long period of operation, while a safe organisation may have accidents. In other words, there is no simple relation between the 'level of safety' of an organisation and the number of adverse outcomes – perhaps because the 'level of safety' is too simple-minded concept.

There is also the problem of whether the measurements refer to snapshots of failures and accidents (cf., Figure 1.1) or whether they relate to the far more frequent acceptable outcomes. A better solution is therefore to base safety management on process measures, where a distinction can be made between direct

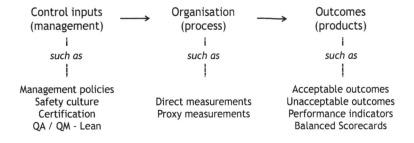

Figure 8.2 Measurement types: product, process, proxy.

measures and proxy measures. If we look for direct measures, which would mean measures related to work as done, there are many. In the case of technological systems, from washing machines to nuclear power plants, it is usually simple both to define relevant process measures (because the processes are designed and well known) and to make the measurements. In the case of sociotechnical systems, it is far more difficult, verging on the impossible. This is not for a lack of trying, such as the various forms of SPC, six sigma, balanced scorecards, etc., illustrate – even though some of them strictly speaking they are outcome measures in disguise. Indeed, it can be a problem that there may be too many direct process measures and that they may be subject to fluctuations or temporary changes. There is a third possibility, namely, to use the four resilience potentials as proxy measures, as discussed in Chapter 5. A proxy measure is an indirect measure that is highly correlated or strongly related to the desired outcome. The desired outcome is in this case resilient performance, as per the definition of resilience.

The purpose of the Resilience Assessment Grid (RAG) is to assess the potentials for resilient performance. It is consistent with the Safety-II perspective that an assessment of the potentials to respond, monitor, learn and anticipate can be used as a 'measure' of safety (or synergy). It is in many ways simpler to assess – or to 'count' – the four potentials than to count specific outcomes in specific categories. Because the assessments have an articulated conceptual basis, they are inherently meaningful. Insofar as the specific items addressed by the RAG give rise to concrete proposals for intervention – keeping in mind the dependencies described by the FRAM model, the RAG can also be used to support the building of a 'culture of resilience'.

The changing face of safety culture

The changing face of safety will inevitably have consequences for a number of other things including the definition of safety culture.

The International Nuclear Safety Advisory Group (INSAG) defined safety culture as 'that assembly of characteristics and attitudes in organizations and individuals which establishes that, as an overriding priority, nuclear plant safety issues receive the attention warranted by their significance' (INSAG, 1991). More recently, it was added that 'the concept of safety culture is a way of exploring how an organization relates to safety issues through a cultural lens' (International Atomic Energy Agency (IAEA), 2016) – although more as a comment than a revision of the original definition. The more popular version is 'the way we do safety around here'. In both cases, the definition of safety culture refers to 'safety', which itself is not defined.

> (This becomes obvious if the definition is rewritten as follows: 'An X culture is that assembly of characteristics and attitudes in organizations and individuals which establishes that, as an overriding priority, X issues receive the attention warranted by their significance'. The ICAO definition

mentioned in Chapter 1 has the same problem: 'A(n) X management system ... is an organized approach to managing X, including the necessary organizational structures, accountabilities, policies and procedures'. In both cases, the definition is analytic, i.e., the concept of the predicate is contained in the concept of the subject – namely 'X'.)

It is clear from the context that the term 'safety' refers to the Safety-I interpretation and that safety therefore is the 'freedom from unacceptable harm'. But there is nothing in the structure of the definitions that prevent them from being used to describe Safety-II or synesis as well. All that is required is to make the predicate more explicit. For Safety-I, the definition could be 'that assembly of characteristics and attitudes in organizations and individuals which contributes to keep the number of adverse outcomes to an acceptably low level'. Likewise, for Safety-II the definition could be 'that assembly of characteristics and attitudes in organizations and individuals which contributes to and sustains successful performance'. In other words, safety culture should not be defined as a 'culture of safety'; it should be defined by something other than itself.

The changing meaning of safety will also have consequences for a number of other concepts that are part of the current Safety-I vocabulary. Reason (1998), for instance, proposed that a safety culture is an informed culture that in turn requires a reporting culture and a just culture. A reporting culture is needed to collect the knowledge gained from incidents, near misses and other 'free lessons', and a just culture is needed to ensure that people are willing to 'confess their own slips, lapses and mistakes'. In both cases, the focus is on adverse events and adverse outcomes, on things that go wrong. If the focus instead is on synesis, on acceptable outcomes and how they are produced, there is little or no need of either a reporting culture or just culture. The reporting is replaced by information about Work-as-Done and about the large and small adjustments that take place throughout an organisation. This information is necessary to manage the organisation, to ensure that it can function as required under expected and unexpected conditions alike, as well as necessary for any learning and improvement to take place. The need of a just culture disappears because people are asked how they do their work, how they cope with unexpected conditions and situations and how they develop robust and effective patterns of work rather than how they have failed. There is no need to offer people protection for sharing their experiences, nor does it usually require much encouragement.

To ensure resilient performance an organisation must know how multiple activities work together to produce acceptable outcomes and how a synthesis can be developed, analysed and sustained. By showing how this can be done in practice, the RAG becomes a tool for the management of Safety-II – for the management of synesis.

A FRAM primer

This Appendix provides a quick introduction, a primer, to the Functional Resonance Analysis Method (FRAM). More extended descriptions can be found on www.functionalresonance.com and of course in Hollnagel (2012).

The purpose of the FRAM is to analyse how something is done, has been done or could be done in order to produce a representation of it. This representation is effectively a model of the activity because it captures the essential features of how something is done, using a well-defined format. In the case of the FRAM, the essential features are the functions that are necessary and sufficient to describe the activity and the way in which they are coupled or mutually dependent.

All safety analysis methods refer to an already existing model. Root Cause Analysis, for instance, represents adverse outcomes and the events that lead to them by single or multiple cause-effect chains starting with the (root) cause and ending with the observed outcomes, cf., the Domino model (Heinrich, 1931). TRIPOD aims to identify the underlying factors of accidents, incidents and near misses as a combination of latent conditions and active failures, cf., the Swiss cheese model (Reason, 1990). The AcciMap approach (Rasmussen and Svedung, 2000) represents accidents in complex sociotechnical system by mapping the possible causes onto six system levels. The map is a network of connections from the physical sequence of events and activities right up to the causes at the governmental, regulatory and societal levels. Each of these methods essentially maps the events onto the model. The same goes for other methods such as STAMP or Bow-Tie.

The FRAM differs in two ways. First, it is not exclusively a safety or risk analysis method but a method to analyse an activity in order to produce a model (description) of it. The FRAM can be used for safety analysis but also for task analysis, system design, etc. Second, it does not refer to an existing model but rather to four principles or assumptions about how things happen. The FRAM is a method-*sine*-model rather than a model-*cum*-method.

First principle: the equivalence of successes and failures

Explanations of accidents and incidents typically rely on decomposing a system or an event into parts, either physical parts such as people and machines or the segments of an activity such as individual actions or steps in a process. Outcomes are explained by linear cause-effect relations among the parts, and adverse outcomes are attributed to malfunctions or failures of parts. This implies that the causes of things that go wrong are different from the causes of things that go right. Otherwise the endeavour to 'find and fix' the causes of unacceptable outcomes would also affect the occurrence of acceptable outcomes. The FRAM – and Resilience Engineering – takes a different approach, namely things that go right and things that go wrong happen in much the same way. The fact that the results are different does not mean that the explanations must be so as well. The principle of approximate adjustments explains why this is so.

Second principle: approximate adjustments

Sociotechnical systems cannot be specified in minute detail because humans are not machines. Effective work requires that performance continuously is adjusted to the existing conditions (resources, time, tools, information, requirements, opportunities, conflicts, interruptions). Adjustments are made by individuals, by groups and by organisations and take place at all levels, from the performance of a specific task to planning and management. Since resources (time, materials, information, etc.) are almost always limited, the adjustments will typically be approximate rather than precise. This is rarely critical because people will know what to expect and be able to compensate for that. The approximate adjustments are the reason why things mostly go right and why they occasionally go wrong.

Third principle: emergent outcomes

The variability of individual functions is rarely large enough to serve as the cause of something going wrong or to be described as a failure. The variability of multiple functions may on the other hand combine in unpredictable (nonlinear) ways that can lead to unexpected and disproportionate outcomes – negative as well as positive. Acceptable and unacceptable outcomes can both be explained as emerging from variability due to the everyday adjustments rather than as a result of single or multiple cause-effect chains starting from a malfunction or failure of a specific component or part.

Fourth principle: functional resonance

As an alternative to linear causality, the FRAM proposes that the variability of two or more functions can coincide and either dampen or amplify each other to produce an outcome or output variability, that is disproportionally large. In the latter case, the consequences may spread to affect other functions in analogy with the phenomenon of resonance.

Functional resonance describes the noticeable performance variability in a sociotechnical system that can happen when multiple approximate adjustments coincide. Performance variability is not random because the approximate adjustments comprise a small set of recognisable adjustments or heuristics. There is some regularity in how people behave and in how they respond to unexpected situations – including those that arise from how other people behave. Functional resonance offers a systematic way to understand outcomes that are both noncausal (emergent) and nonlinear (disproportionate).

Basic concepts in developing a FRAM model

The FRAM is a systematic approach to creating a description or representation of how an activity usually takes place. This representation is called a FRAM model. The performance is described in terms of the functions that are necessary to carry out the activity, the potential couplings between the functions and the typical variability of the functions. The purpose of the FRAM is to provide a concise and systematic description of work as it typically takes place.

The meaning of functions in the FRAM

A function in the FRAM represents the means that are necessary to achieve a goal. More generally, a function represents the acts or activities – simple or complicated – that are needed to produce a certain result.

- A function typically describes what people – individually or collectively – have to do to perform a specific task and thus achieve a specific goal, for example, triage a patient or guide an approaching aircraft.
- A function can also refer to something that an organisation does: for example, it is the function of a railway line to transport people and goods.
- A function can finally refer to what a technical system does either by itself (an automated function, such as a robot) or together with one or more people (an interactive or sociotechnical function, like a check-in kiosk in an airport).

To emphasise that functions represent activities or something that is done, it is recommended to describe them by a verb or a verb phrase. For instance, '(to) diagnose a patient' rather than 'diagnosing a patient or '(to) request information' rather than 'requesting information'.

The meaning of aspects in the FRAM

Functions can be described in terms of six aspects: Input, Output, Requirements, Resources, Control and Time. The general rule of the FRAM is that a function's aspects should be described if the analysis team thinks it is appropriate and if there is sufficient information or experience to do so. It is not necessary to describe all six aspects of every function, and it can indeed sometimes be either impossible or unreasonable to do so. As a minimum, at least one Input

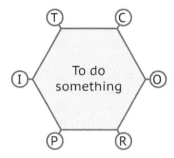

Figure A.1 A FRAM function.

and one Output must be described for all foreground functions. Note, however, that a FRAM model is reduced to an ordinary flow chart or network diagram if only the Input and Output aspects are described. The FRAM recommends that an aspect be described with a noun or a noun phrase. In other words, an aspect is described as a state or as a result of something – but not as an activity (Figure A.1).

- The Input (I) to a function is traditionally used or transformed by the function to produce the Output. The Input can represent matter, energy or information. An Input can also be what activates or starts a function, such as a clearance or an instruction to begin doing something. Input can be seen as a form of data or information or more generally as something that a function interprets as a signal to begin. Formally, an Input is always the result of a change in the state of the environment whether in terms of energy, information or position. For that reason, the description of the Input is always a noun or a noun phrase. When an Input aspect is described for one function, it must also be described as an Output from another function.
- The Output (O) of a function describes the result of what the function does, for example, the result of processing the Input. The Output may represent material, energy or information – an example of the latter would be a permission or clearance or the result of a decision. The Output describes a change of state of one or more output parameters. The Output may be, for example, the signal to start another function. The description of the Output should be a noun or a noun phrase. When an Output aspect is described for one function, it must also be defined as an Input, Precondition, Resource, Control or Time aspect for another function.
- Some functions cannot begin before one or more Preconditions (P) have been established. These Preconditions can be understood as system states that must exist, or as conditions that must be verified before a function is carried out. A Precondition, however, does not itself constitute a signal

that can activate a function. This simple rule can be used to determine whether something should be described as an Input or as a Precondition. The description of a Precondition should be a noun or a noun phrase. When a Precondition aspect is defined for one function, it must also be defined as an Output from another function.

- A Resource (R) is something that is needed or consumed while a function is carried out. A Resource can represent matter, energy, information, competence, software, tools, manpower, etc. Since some resources are consumed while a function is carried out while others are not, it is useful to distinguish between (proper) Resources on the one hand and Execution Conditions on the other. A (proper) Resource is consumed by a function and will therefore be gradually reduced; an Execution Condition only needs to be available or exist while a function is active. (The difference between a Precondition and an Execution Condition is that the former is only required before a function starts but not while it is carried out.) The description of a Resource (an Execution Condition) should be a noun or a noun phrase. When a Resource aspect is defined for one function, it must also be defined as an Output from another function.

- Control (C) is that which regulates a function so that it produces the desired Output. Control can be a plan, a schedule, a procedure, a set of guidelines or instructions, a program (an algorithm), a 'measure and correct' functionality, etc. Another, less formal type of control is social control or expectations of how the work should be done. Social control can be external, such as the expectations of others (management, organisation, co-workers). Social control can also be internal, for example, a habitual plan of action for a job or what we imagine others expect from us. The description of Control should be a noun or a noun phrase. When a Control aspect is defined for one function, it must also be defined as an Output from another function.

- Time (T) represents the various ways in which time can affect the performance of a function. A function may, for instance, have to be carried out (or be completed) before another function, after another function, overlapping with – parallel to – another function or within a certain duration. The description of a Time aspect should be a noun, if it is a single word, or begin with a noun, if it is a short sentence. When a Time aspect is defined for one function, it must also be defined as an Output from another function.

Couplings

Couplings describe how functions are connected or depend upon each other. Formally, two functions are said to be coupled if an Output from one corresponds to an Input, Precondition, Resource, Control or Time of another. The couplings that are described by a FRAM model, i.e., the dependencies that are the due to common aspects, are called potential couplings because a FRAM model describes the possible relationships or dependencies between functions without referring to a particular situation. An instantiation of a FRAM model represents how a subset of functions can actually become coupled under given conditions or within a given time frame. The subset represents the actual couplings or dependencies that have occurred or are expected to occur in a particular situation or a particular scenario. The couplings described for a specific instantiation do not change but are 'fixed' or 'frozen' for the assumed conditions. For an event analysis, the instantiation will typically cover the entire duration of the event and the couplings that existed then.

Foreground and background functions

Functions in the FRAM can be described as either foreground functions or background functions. The terms have nothing to do with the type of functions that are involved but with the role of the function in a particular model – and of course also for instantiations of the model. A function is considered as a foreground function if it is part of the study focus, which in practice means if the variability of the function may have consequences for the outcome of the activity being examined. A background function is used to account for something that is used by foreground functions, but which is assumed to be stable in the situation under consideration. It could, for example, be a Resource (the right level of staffing or the competence of the staff) or an instruction (Control). A person's competence must generally be assumed to be stable (not varying) during the execution of a task, just as an instruction also must be assumed to be stable. This does not mean that competence is sufficient or that the instruction is correct, but only that it should be regarded as stable during the time it takes to perform the task. Foreground and background function thus refer to the relative importance of a function in the model and not to a function as such. If the study focus changes, a function may change from being a foreground function and become a background function and vice versa.

For foreground functions, it is necessary to describe at least Input and Output. For background functions that represent a source of something, it may be sufficient to describe the Output. Similarly, for background functions that serve as placeholders for downstream functions not included in the analysis (i.e., as drains), it may be sufficient to describe the Input. This means that the expansion of a FRAM model stops whenever a background function is reached. (Figure 6.5, for instance, contains nine background functions but only two foreground functions.)

Upstream and downstream functions

While the terms foreground and background represent a function's role in a model, the terms upstream and downstream are used to describe the temporal relationship between a function that is in focus and the other functions. The analysis of the FRAM model takes place by following the potential couplings between the functions step by step. This means that there will always be one or more functions in focus, i.e., whose variability is being considered. The functions that have been in focus before, meaning functions that already have been performed, are referred to as upstream functions. Similarly, the functions that follow the function that is in focus are called downstream functions. During the implementation of an analysis, any function can change status from being downstream, to come into focus, to become an upstream function.

A FRAM model describes the functions and their potential couplings for a typical situation but not for a specific situation. It is therefore not possible to say with certainty whether a function always will be performed before or after another function. That can only be determined when the model is instantiated. By contrast, the labels foreground function and background function are valid both for the FRAM model as its instantiations. An instantiation of the model uses detailed information about a particular situation or scenario to create an instance or a specific example of the model. This corresponds to a temporal organisation of functions that reflects the order in which they will take place in the scenario, depending on how much variability there is in each function, in the operating environment and in the upstream-downstream couplings.

Graphical representation of a FRAM model

As explained above, a FRAM model represents a system's functions (the union of the foreground and background functions). The model also describes the potential couplings between the functions that can be derived from the functions' aspects. A graphical representation of a FRAM model uses hexagons to represent functions and shows the potential connections between the functions. (The FRAM Model Visualiser or FMV has been used for all FRAM models shown in this book. The FMV is not described here but can be found at www.functionalresonance.com, from which the current version, as well as a brief set of instructions, can be downloaded.) The graphical representation does not define a default orientation or ordering of the hexagons (such as from left to right or from top to bottom).

References

Ashby, W. R. (1956). *An introduction to cybernetics.* London: Chapman & Hall.

Australian Radiation Protection and Nuclear Safety Agency (ARPANSA) (2012). Holistic Safety Guidelines V1 (OS-LA-SUP-240U). Melbourne, Australia: ARPANSA.

Baumard, P. and Starbuck, W. J. (2005). Learning from failures: Why it may not happen. *Long Range Planning, 38,* 281–298.

Besnard, D. and Hollnagel, E. (2012). I want to believe: Some myths about the management of industrial safety. *Cognition, Technology & Work, 16*(1), 13–23.

Burke, W. W. and Litwin, G. H. (1992). A causal model of organizational performance and change. *Journal of Management, 18*(3), 523–545.

Carpenter, S. et al. (2001). From metaphor to measurement: Resilience of what to what? *Ecosystems, 4,* 765–781.

Chapman, D. W. and Volkman, J. (1939). A social determinant of the level of aspiration. *Journal of Abnormal and Social Psychology, 34,* 225–238.

Conant, R. C. and Ashby, W. R. (1970). Every good regulator of a system must be a model of that system. *International Journal of Systems Science, 1*(2), 89–97.

Dekker, S. W. A. and Hollnagel, E. (2004). Human factors and folk models. *Cognition, Technology & Work, 6,* 79–86.

Foster, P. and Hoult, S. (2013). The safety journey: Using a safety maturity model for safety planning and assurance in the UK Coal Mining Industry. *Minerals, 3,* 59–72.

Haavik, T. K. et al. (2016). HRO and RE: A pragmatic perspective. *Safety Science,* http://dx.doi.org/10.1016/j.ssci.2016.08.010.

Hale, A. R. and Hovden, J. (1998). Management and culture: The third age of safety. A review of approaches to organizational aspects of safety, health and environment. In A. M. Feyer and A. Williamson (Eds.), *Occupational injury. Risk prevention and intervention.* London: Taylor & Francis.

Hamel, G. and Välikangas, L. (2003). The quest for resilience. *Harvard Business Review, 81*(9), 52–65.

Heinrich, H. W. (1931). *Industrial accident prevention.* New York: McGraw-Hill.

Holling, C. S. (1973). Resilience and stability of ecological systems. *Annual Review of Ecology and Systematics, 4,* 1–23.

Hollnagel, E. (2001). "Managing the Risks of Organizational Accidents" from the cognitive systems engineering viewpoint. Presentation at panel discussion on the "Prevention and Risk-mitigation of System Accidents from the Human-Machine Systems (HMS) Viewpoint". 8th IFAC/IFIP/IFORS/IEA *Symposium on Analysis, Design, and Evaluation of Human–Machine Systems,* Kassel, Germany, 18–20 September.

Hollnagel, E. (2006). Resilience – the challenge of the unstable. In E. Hollnagel, D. D. Woods and N. C. Leveson (Eds.), *Resilience engineering: Concepts and precepts*. Aldershot, UK: Ashgate.

Hollnagel, E. (2009a). *The ETTO principle: Efficiency-thoroughness trade-off. Why things that go right sometimes go wrong*. Farnham, UK: Ashgate.

Hollnagel, E. (2009b). The four cornerstones of resilience engineering. In C. P. Nemeth, E. Hollnagel and Dekker, S. (Eds.), *Preparation and restoration* (pp. 117–134). Aldershot, UK: Ashgate.

Hollnagel, E. (2011). Prologue: The scope of resilience engineering. In E. Hollnagel et al. (Eds). *Resilience engineering in practice. A guidebook*. Farnham, UK: Ashgate.

Hollnagel, E. (2012). *FRAM – the functional resonance analysis method: Modelling complex socio-technical systems*. Farnham, UK: Ashgate.

Hollnagel, E. (2014a). *Safety-I and Safety-II: The past and future of safety management*. Farnham, UK: Ashgate.

Hollnagel, E. (2014b). Is safety a subject for science? *Safety Science*, 67, 21–24.

Hunte, G. and Marsden, J. (2016). *Engineering resilience in an urban emergency department, Part 2*. Paper presented at the Fifth Resilient Health Care Meeting, August 15–17, Middelfart, Denmark. http://resilienthealthcare.net/meetings/denmark%202016.html.

ICAO (2006). *Safety Management Manual* (SMM) (DOC 9859 AN/460). Montreal, Canada: International Civil Aviation Organization.

International Atomic Energy Agency (IAEA). (2016). *Performing safety culture self-assessments*. Wien, Austria: International Atomic Energy Agency.

International Nuclear Safety Advisory Group (INSAG). (1991). *Safety culture*. Wien, Austria: International Atomic Energy Agency.

Kaplan, R. S. and Norton, D. P. (1992). The balanced scorecard – measures that drive performance. *Harvard Business Review, January–February*, 71–79.

Keesing, R. M. (1974). Theories of culture. *Annual Review of Anthropology*, 3, 73–97.

Kletz, T. (1994). *Learning from accidents*. London: Butterworth-Heinemann.

Ljungberg, D. and Lundh, V. (2013). *Resilience Engineering within ATM – development, adaption, and application of the Resilience Analysis Grid (RAG)* (LiU-ITN-TEK-G—013/080—SE). Linköping, Sweden: University of Linköping.

March, J. G. (1991). Exploration and exploitation in organizational learning. *Organization Science*, 2(1), 71–87.

Maslow, A. H. (1943). A theory of human motivation. *Psychological Review*, 50, 370–396.

Maslow, A. H. (1965). *Eupsychian management*. Homewood, IL: Richard D. Irwin/The Dorsey Press.

McGregor, D. (1960). *The human side of enterprise*. New York: McGraw-Hill.

Miller, J. G. (1960). Information input overload and psychopathology. *American Journal of Psychiatry*, 116, 695–704.

Miller, G. A., Galanter, E. and Pribram, K. H. (1960). *Plans and the structure of behavior*. New York: Holt, Rinehart & Winston.

Moon, S. et al. (2015). Will Ebola change the game? Ten essential reforms before the next pandemic. The report of the Harvard-LSHTM Independent Panel on the Global Response to Ebola. *The Lancet*, 386(10009), 2204–2221.

Parker, D., Lawrie, M. and Hudson, P. (2006). A framework for understanding the development of organisational safety culture. *Safety Science*, 44, 551–562.

Perrow, C. (1984). *Normal accidents*. New York: Basic Books.

Pringle, J. W. S. (1951). On the parallel between learning and evolution. *Behaviour, 3,* 175–215.

Prochaska, J. O. and DiClemente, C. C. (1983). Stages and processes of self-change of smoking: Toward an integrative model of change. *Journal of Consulting and Clinical Psychology, 51*(3), 390–395.

Rasmussen, J. and Svedung, I. (2000). *Proactive risk management in a dynamic society.* Karlstad, Sweden: Swedish Rescue Services Agency.

Reason, J. T. (1990). The contribution of latent human failures to the breakdown of complex systems. *Philosophical Transactions of the Royal Society (London), Series B. 327,* 475–484.

Reason, J. T. (1998). Achieving a safe culture: theory and practice. *Work & Stress, 12*(3), 293–306.

Reason, J. T. (2000). Safety paradoxes and safety culture. *Injury Control & Safety Promotion, 7*(1), 3–14.

Rigaud, E. et al. (2013). Proposition of an organisational resilience assessment framework dedicated to railway traffic management. In N. Dadashi et al. (Eds.), *Rail human factors: Supporting reliability, safety and cost reduction.* London: Taylor & Francis.

Schein, E. H. (1990). Organisational culture. *American Psychologist, 45*(2), 109–119.

Shewhart, W. A. (1931). *Economic control of quality on manufactured product.* New York: D. Van Nostrand Company.

Taylor, F. W. (1911). *The principles of scientific management.* New York: Harper.

Tredgold, T. (1818). On the transverse strength of timber. *Philosophical Magazine: A Journal of Theoretical, Experimental and Applied Science,* Chapter XXXXVII. London: Taylor and Francis.

Tuchman, B. W. (1985). *The march of folly: From Troy to Vietnam.* New York: Ballantine Books.

VMIA (2010). *Reducing harm in blood transfusion. Investigating the Human Factors behind 'Wrong Blood in Tube'* (WBIT). Melbourne, Australia: Victoria Managed Insurance Authority.

Weick, K. E. (1987). Organizational culture as a source of high reliability. *California Management Review, 29*(2), 112–128.

Weinberg, G. M. and Weinberg, D. (1979). *On the design of stable systems.* New York: Wiley.

Westrum, R. (1993). Cultures with requisite imagination. In J. A. Wise, V. D. Hopkin and P. Stager (Eds.), *Verification ad validation of complex systems: Human factors issues* (pp. 401–416). Berlin: Springer Verlag.

Westrum, R. (2006). A typology of resilience situations. In E. Hollnagel, D. D. Woods and N. Leveson (Eds.) *Resilience engineering. Concepts and precepts.* Ashgate: Aldershot, UK.

Wiener, N. (1954). *The human use of human beings.* Boston, MA: Houghton Mifflin Co.

Woods, D. D. (2000). *Designing for resilience in the face of change and surprise: Creating safety under pressure.* Plenary Talk, Design for Safety Workshop, NASA Ames Research Center, October 10.

Glossary

Causality credo. Within the Safety-I tradition, explanations of how accidents happen share the unspoken assumption that outcomes can be understood as effects that follow from prior causes. Since that corresponds to a belief – or even a faith – in the law of causality, it may be called a causality credo. Reasoning according to the causality credo goes through the following steps: (1) adverse outcomes happen because something has gone wrong; (2) if enough evidence is collected, it will be possible to find the causes and then eliminate, encapsulate or otherwise neutralise them and (3) since all adverse outcomes have causes, and since all causes can be found and dealt with, it follows that all accidents can be prevented. The Zero Accident Vision is thus a logical consequence of the causality credo.

Complexity. The term 'complexity', either used alone or as an adjective such as in 'complex adaptive system', is an example of a monolithic explanation (*q. v.*). It is in many cases impossible to have complete knowledge of the systems that we must manage and of the organisations within which we function. This is explained by pointing out that the systems and organisations are complex. But complexity refers as much to the human inability to understand something (a phenomenon or a target system) as to a quality or characteristic of the system *eo ipso*. It is therefore unwarranted to claim that complexity is ontological rather than epistemic.

Culture represents the common agreement, sometimes explicit but usually tacit, in a group of people, sometimes small but usually very large, on how to behave and what to do both generally and in specific conditions. As such it is an important determinant of both individual and collective behaviour, although it is not the only one. Culture is often used as an unspecified suffix, e.g., safety culture, just culture, reporting culture.

Hypothesis of different causes. According to the hypothesis of different causes, the causes of things that go wrong (accidents, incidents) are different from the causes of things that go right (acceptable outcomes). If that was not so, then the 'find-and-fix' solution – eliminating the causes of things that go wrong – would also affect the causes of things that go right. Although the hypothesis rarely is stated explicitly, it an inherent part of Safety-I.

Intractable. A system is said to be intractable if the principles of its functioning only are partly known (or, in extreme cases, completely unknown), if the descriptions of it are elaborate with many details or if the system is unstable or develop so rapidly that it changes faster than a description can be made.

Law of Requisite Variety. This law, which is part of cybernetics – the science of "control and communication in the animal and the machine" – states that in order to ensure a sufficiently small variety of outcomes from a system it is necessary that the regulator of that system has as much variety itself. Simply put, if more things can happen in an organisation than the leaders or management can imagine (and prepare for), then the organisation cannot be effectively managed.

Monolithic explanations. Explanations and/or descriptions of nontrivial, dynamic events often use a single concept or factor to 'solve' a host of problems. Such many-to-one solutions are obviously attractive since they make it easier both to explain something and to communicate it to others. Since such explanations rely on a single concept they can be called monolithic. While monolithic explanations are psychologically satisfying, they have limited practical value. Monolithic explanations represent a social convention and are therefore essentially social constructs.

Nontrivial. A nontrivial system is partly and irreducibly unpredictable. This means that it is uncertain how the system will develop, even if left on its own, and uncertain what the response to changes and interventions will be. The lack of predictability is due to incomplete or insufficient knowledge about the system and how it functions.

Resilience. The performance of a system is resilient if it can function as required under expected and unexpected conditions alike (changes/disturbances/opportunities). Resilient performance requires that an organisation has the potentials for resilience and that these potentials are continuously developed, maintained and improved. Since resilience is a manifestation of the resilience potentials, it refers to something that an organisation does, rather than to something that it has.

Safety is usually defined in terms of the absence of or freedom from something, as being *without* unacceptable or adverse outcomes. This leads to the paradoxical condition that safety is defined by situations or conditions where it is absent rather than present.

Safety-I is a condition where the number of unacceptable outcomes (accidents/incidents/near misses) is as low as possible. Safety-I is therefore defined by its opposite – by the lack of safety (accidents, incidents, risks).

Safety-II is a condition where the number of acceptable outcomes (meaning everyday work) is as high as possible. Safety-II is therefore defined by the presence of safety – as synesis (*q. v.*).

Safety culture is an example of a monolithic explanation (*q. v.*) where culture is used as an unspecified suffix. The standard definition of *safety* culture as "that assembly of characteristics and attitudes in organisations and individuals which establishes that, as an overriding priority, nuclear plant *safety* issues receive the attention warranted by their significance" is furthermore self-referential, hence useless unless it is mutually agreed what 'safety' is.

Synesis. A condition where multiple activities fit together to bring about the intended outcome in an effective way. The mutually dependent set of functions that is necessary for a workplace to carry out its activities as intended with respect to a set of relevant criteria (safety, quality, productivity, etc.).

System. A system is usually defined in terms of its structure as "a set of objects together with relationships between the objects and between their attributes". But a system may also be defined as the set of coupled functions that are needed to provide a certain performance. An organisation is a system in both senses, but the functional definition is both more interesting and more useful than the structural definition.

Tractable. A system is tractable if the principles of its functioning are known, if descriptions of it are simple and with few details and, most importantly, if it does not change during the time it takes to describe it.

Trivial. Something, a system or a development, is called trivial if it is evident what is going to happen. A trivial system is first and foremost predictable. The predictability means that the way the system develops or changes when it is on its own (*eigenleben*) can be foreseen. The predictability also means that the system's responses to changes and interventions (managed inputs) can be predicted. In either case the predictions can be made with high certainty in the short term and with lower, but still acceptable certainty, in the longer or long term. The second form of predictability is necessary in order for the system to be controllable (cf., the Law of Requisite Variety).

Index

Milton Keynes UK
Ingram Content Group UK Ltd.
UKHW022038141024
449569UK00014B/651